Legal information

Cover image: Designed by macrovector / Freepik (modified)

© 2023

Author and Editor: M.Eng. Johannes Wild

A94689H39927F

Email: 3dtech@gmx.de

The complete imprint of the book can be found on the last pages!

This work is protected by copyright

Thank you so much for choosing this book!

Table of Contents

Introduction

Thank you very much for choosing this book!

As an engineer and 3D printing enthusiast, I am familiar with all the pitfalls of 3D printing in detail. 3D printing is a process that needs to be mastered. The 3D Printer may initially (and this is what most people – including me – have experienced at the beginning) produce a variety of unsatisfactory results.

This can literally drive you crazy and make you want to give up on this ingenious technology. Especially if the poor printing results are not limited to the beginning of your printing, but occur again and again, with no solution at hand.

This is over now! For all those who are looking for a handbook of all 3D printing failures, this compact manual was written. In the following chapters, you will find a description of each issue pattern, followed by a section about the possible causes and specific steps to solve the problem. Numerous images help you with troubleshooting your way through 3D Printing.

This book is designed for both beginners and intermediates.

With this handbook, you can gain a solid understanding of solving 3D printing trouble and learn everything you need to know to:

a) avoid 3D printing errors from the very beginning,

b) classify 3D printing failures that have already occurred,

c) understand the underlying causes, and,

d) take an appropriate troubleshooting approach

That being said, let's get started!

1 Basics: Tips & Tricks for High-Quality Printing Results

Before you start a 3D printing job, you should ensure that the following basics have been considered sufficiently:

a) The print bed is leveled: Print bed leveling is one of the most important procedures in FDM 3D printing. If the distance between the nozzle and the print bed is not correct (too large / too small), the object will not adhere. In chapter 4, you will find a leveling guide that explains the leveling process in detail.

b) Bolts are tightened / belts are tensioned: Check all bolts of the 3D Printer for sufficient tightness and whether all drive belts are sufficiently tensioned because mechanical discrepancies can also lead to printing errors.

c) Solid installation site: The 3D Printer should be placed on a solid and vibration-dampening platform, such as a solid workbench or aluminum frame. Otherwise, vibrations may build up and be transmitted to the print head, and thus affecting the print result negatively.

d) Environment: Operate the 3D Printer in a room with a constant temperature in the range of 20 - 23° C (65 - 75° F) and a low humidity level (up to about 40 %). Opened filament must be stored in a closed box with moisture-absorbing material (e.g., silica gel). The room should also be as clean as possible. Avoid any excessive amounts of dust, oil, or other substances (garage (oil), laundry room (humidity): rather unsuitable).

e) Printing Temperature: Choose a print bed & nozzle temperature that is suitable for your filament. This is usually specified by the filament manufacturer itself and can be found on the packaging. Print a "Temperature Tower" (thingiverse.com) in order to find the optimal printing temperature.

f) Print speed: Print at a slow (overall) speed (e.g.: 60 - 100 mm/s). This will give you a high print quality.

Safety Instructions for 3D Printing:

Figure 1: Safety Instructions

The printing nozzle must heat up to approx. 200° C (400° F) to melt the filament. The print bed also heats up to approx. 60° C (140° F). Therefore, there is a high risk of burns, especially at the nozzle. Do not touch moving parts, either; while the printer is running, there is a danger of crushing. Find a suitable place to set up the 3D Printer, ideally in a somewhat closed room, as the printer does generate noise and vapors (hazardous to health, e.g., with ABS). Also, make sure that children and pets do not have access to the 3D Printer. It is also a good idea to spend a few minutes at the start of each print job to check whether everything is being carried out correctly. In addition, you could install a camera for remote monitoring.

In the next chapter, you will find an image-based directory containing chapter references for easy identification of printing failures (compare/search for your pattern and look up the chapter reference), before we deal with the specific error patterns in detail. In addition, a chapter on critical components of an FDM 3D Printer and a leveling guide is waiting for you!

2 Image-based directory for easy troubleshooting

9.14
9.11
9.22
9.20
9.16
9.12
9.19
9.21
9.13
9.9
10.1
9.18
9.8

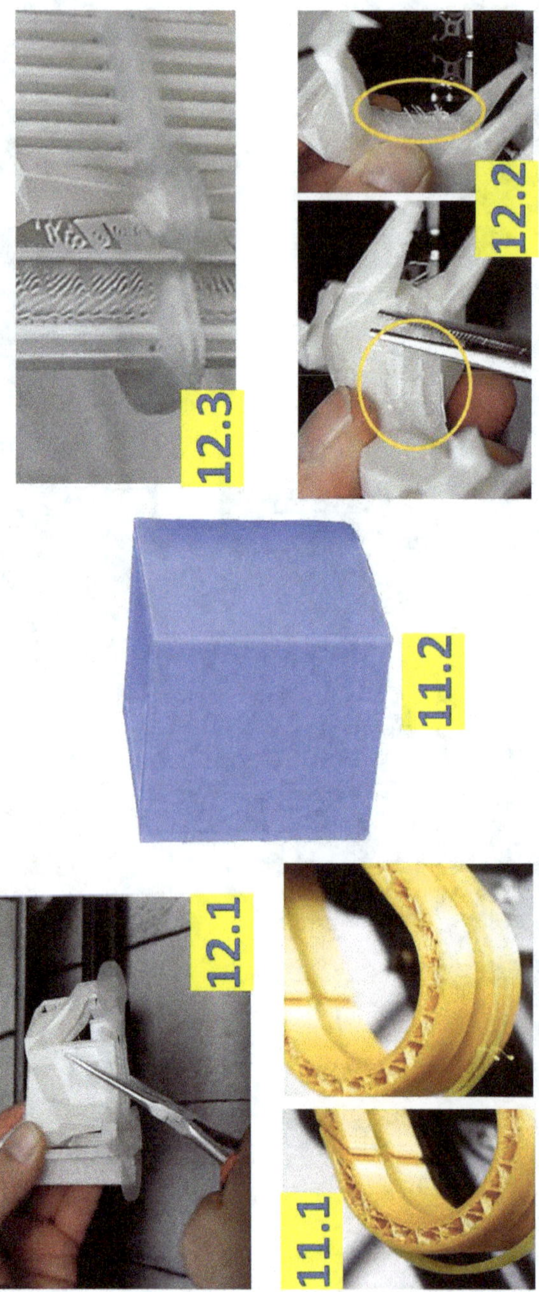
12.3
12.2
11.2
12.1
11.1

3 Critical Components of an FDM 3D Printer

An FDM 3D Printer consists of the following core elements:

i. Frame for holding the components,
ii. Print platform (or print bed) with a heating element (optional) and adjustment features for height regulation,
iii. Stepper motors with drive elements: pulley/belt,
iv. Extruder for filament feed,
v. Motorized z-axis(es),
vi. Hot-End with nozzle and fans,
vii. Electronics with controller & power supply.

optional: filament sensor (stops the printer in case of faulty filament supply), Auto-Bed-Leveling Sensor (for automatic bed leveling), **permanent printing bed (a mirror is used here; "Buildtak" is also recommended)**

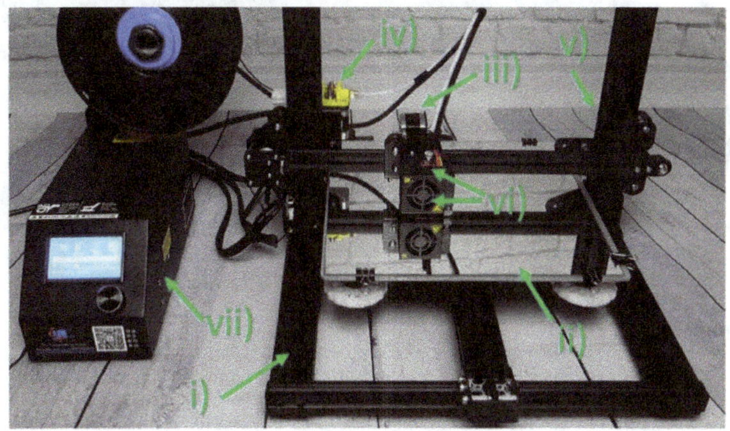

Figure 2: Components of an FDM 3D Printer (Creality's CR-10)

Critical Components for Print Quality Issues:

From a mechanical perspective, the overall stiffness of the frame (struts / mechanical connections), the drive elements (belts), guide elements (ball bearings/pulleys), and the z-axis(es) are particularly critical for the achievable print quality. Therefore, a printer with a rigid frame should be selected (if necessary, struts should be attached), all drive belts should be sufficiently tensioned, mechanical connections should be checked, and the guide elements should be checked for adequate tolerances (not too much / not too little).

Other critical components are the heating elements, such as the hot end and the heating element for the printing bed, as well as their PID controller (closed-loop control). These elements should be able to provide a very constant heating power throughout the printing process. Too high fluctuations can lead to problems, especially in the area of the hot end and the nozzle. In most cases, however, the PID controller works reliably.

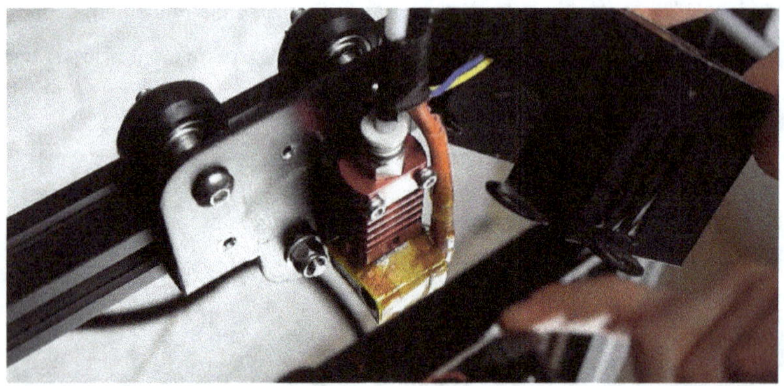

Figure 3: The Hot End of a 3D Printer (CR-10)

The main source of 3D printing issues (not to mention a poorly leveled printing bed, which is actually one of the major failures of 3D printing beginners) is found in the slicing settings. Therefore, the choice of these settings has a fundamental impact on the quality of the printed object. For this reason, the following chapters provide explicit advice and recommendations on slicing settings. The recommended slicing software is Cura by Ultimaker, which is available for free.

A mirror is recommended as a permanent printing bed. Mirrors have to be manufactured extremely plane in order not to reflect a distorted image, and therefore represent an ideal low-cost platform for 3D printing. Using a mirror, the adhesion of the object is surprisingly good, and the bottom of the print gets very smooth. 3D Printers for less than $ 500 usually have a lower manufacturing accuracy of elements such as the printing bed (aluminum or glass plate), which can cause the printing bed to "sag," especially in the middle area, and thus making even the most careful printing bed leveling pointless.

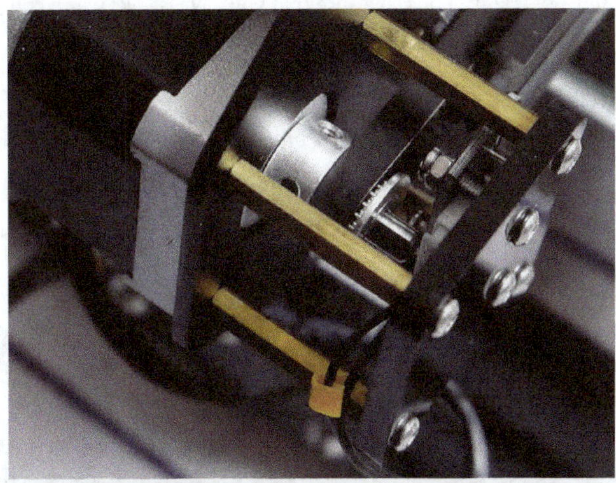

Figure 4: Stepper motors with pulley, guiding, and drive belt

4 The Alpha and Omega: Leveling Guide

For the process of manual leveling, heat up the nozzle and the printing bed to an adequate temperature via the menu of the 3D Printer (nozzle: 210° C (400° F); printing bed: 60° C (140° F)). Make sure that you always level warm components, as the materials expand under the influence of heat and thus would falsify the result. Then select "Home Position" under "Prepare" in the printer's menu (example: CR-10). The print head will then move to a defined home position, in this case, to the front left corner of the print bed. This is necessary to always achieve the same z-distance between nozzle and printing bed at the beginning of a printing process.

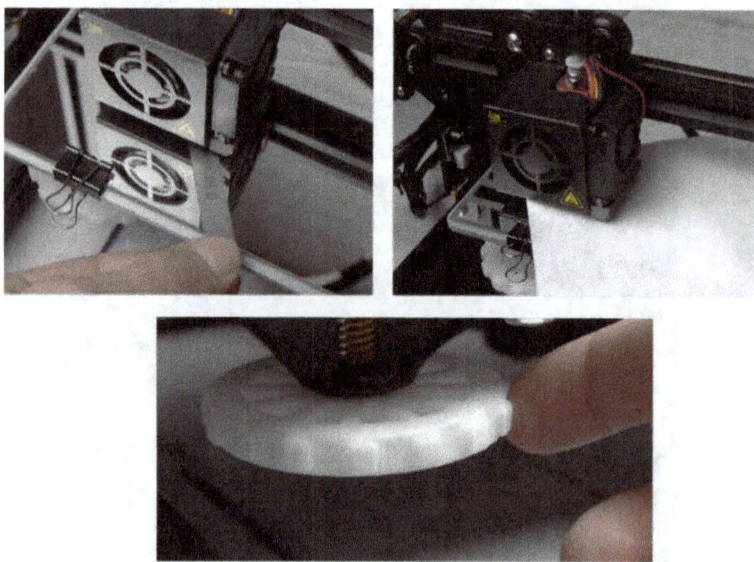

Figure 5: Level the printing bed using a distance gauge or a sheet of paper

The leveling process works best with a feeler gauge of 0.1 or 0.2 mm. If you do not have one at hand, you can alternatively use a piece of paper. Place the gauge or paper between the nozzle and the printing bed to check the distance. The gauge should just fit between the nozzle and the printing bed. When using a piece of paper, you should hear a slight (!) scraping noise. If the distance is too big or too small, adjust the distance with the wheels of the printing bed.

Repeat this procedure in all four corners and redo it once or even twice (until you feel confident that the distance is consistent). If you are using a sheet of paper, make sure that the sheet can only be moved very slightly. Then the gap is just right.

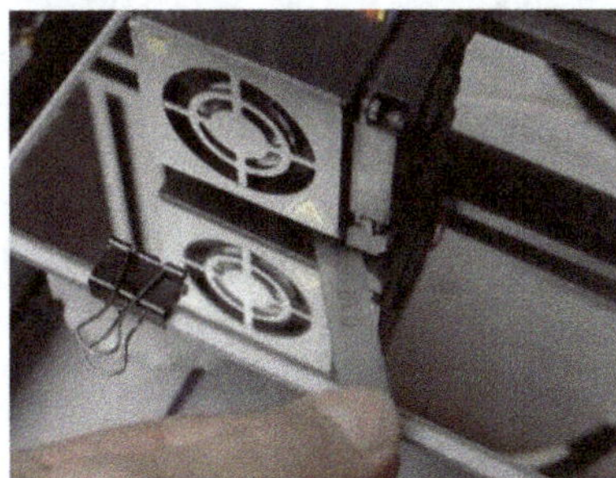

Figure 6: Always level the print bed at several points

If the first print does not adhere, it is advisable to reduce the distance between the nozzle and the bed slightly. A too-large distance between nozzle and print bed is one of the most common beginner's mistakes. On the other hand, the nozzle should never grind on the print bed while the print head is moving! In the worst case, this can damage your print bed, your print nozzle, or other parts of the printer.

Recent 3D Printers – depending on the model – often already have a sensor for automatic regulation of the distance between the printing nozzle and the printing bed. But it does make sense to carry out a manual leveling during the first set-up to obtain a well-prepared starting surface.

For printers without a sensor, such as the first CR-10, print bed leveling should be carried out regularly each 10 - 20 prints or in case of other discrepancies such as poor adhesion. For printers with sensors, such as the CR-10s Pro, performing this manual process on a regular basis is generally no longer necessary.

5 General Issues concerning the Print Bed

5.1 The Printing Bed does not (fully) reach the set Temperature

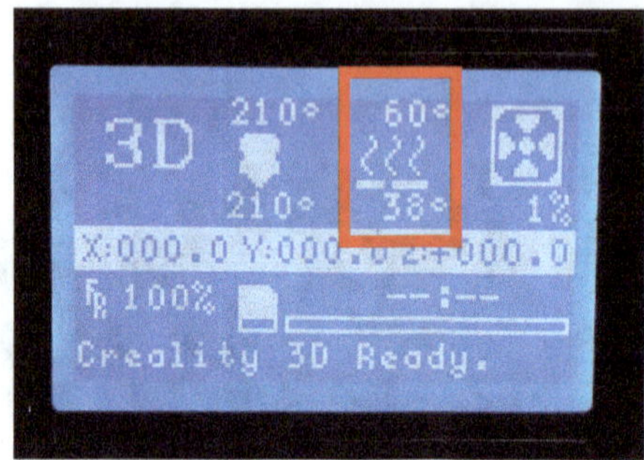

Figure 7: The Printing bed does not reach the set temperature

Issue:

The printing bed remains cold or does not reach the set temperature.

Cause(s):

- i. The heating bed cable or the connecting plug is broken/loose.
- ii. The "gcode" was created without settings for the printing bed temperature.
- iii. The 3D Printer's PID-controller is broken.
- iv. Low room temperature; it takes a very long time for the printer to heat up / to reach the final temperature.

Solution(s):

- i. Replace the heating bed cable and check all connections using a multimeter.
- ii. Check the slicing settings for the correct printing bed temperature and create a new "gcode."
- iii. Replace the PID controller or have it replaced by a professional.
- iv. Use a housing for the 3D Printer (heated if necessary) and / or print in a room with at least 22° C (70° F).

5.2 Printing Bed Leveling is not good

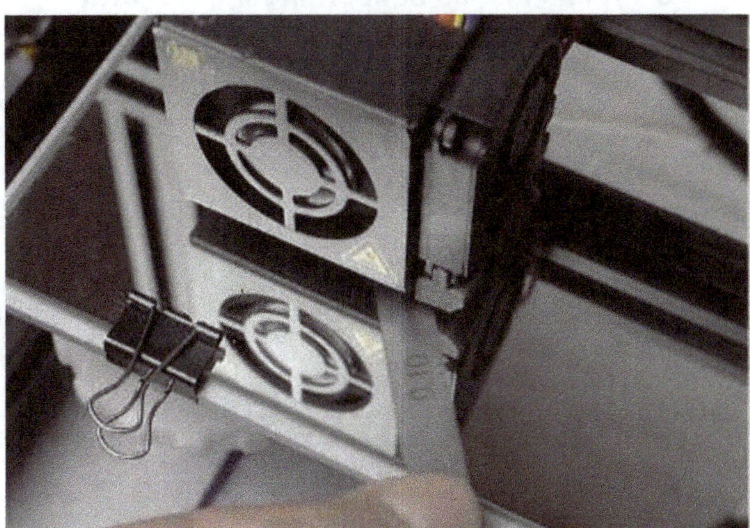

Figure 8: The Leveling of the printing bed fails

Issue:

The printing bed does not become plane enough despite repeated leveling; the distance between nozzle and printing bed is not correct in some areas.

Cause(s):

  i.   The printing bed platform is not manufactured plane enough.
  ii.   The printing bed platform bulges in the middle ("sagging").
  iii. The axis of the 3D Printer is/are not aligned parallel to the printing bed.

Solution(s):

  i.   Replace the printing bed platform / use a mirror (see Leveling Guide).
  ii.   Mount a small adjustable lifting jack (can be 3D printed) beneath the center of the printing bed to compensate any bulging.
  iii. Check the axes of the 3D Printer for proper mounting and alignment.

6 Typical Issues concerning the Hot End

6.1 The Hot End does not heat

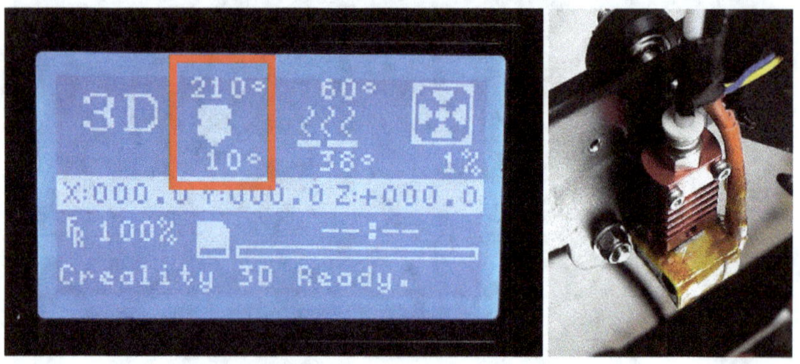

Figure 9: The nozzle does not reach the set temperature

Issue:

The hot end remains cold or does not reach the set temperature.

Cause(s):

i. A loose contact of the connecting plugs or a broken hot end wiring.
ii. The "gcode" was created incorrectly.
iii. The 3D Printer's PID controller is defective.
iv. The room temperature is very low; it takes a long time for the printer to heat up.
v. If the hot end does not reach the final temperature: Try slowing down the fans (slicing settings).

Solution(s):

i. Replace the heating bed cable and check all connections using a multimeter.
ii. Check the slicing settings for the correct printing bed temperature and slice again.
iii. Replace the PID controller or get it done by a professional.
iv. Use a (heated) housing for the 3D Printer.
v. Slow down the fans (slicing settings).

6.2 Clogged Nozzle

Figure 10: The nozzle of the 3D Printer is clogged

Issue:

The nozzle of the 3D Printer emits no material or only minimal amounts. You may also hear clicking noises from the extruder.

Cause(s):

i. There are filament residues in the nozzle.
ii. A feed rate that is too fast and thus resulting in a material jam.
iii. The outlet of the nozzle is obstructed (wear/deformation due to mechanical damage).
iv. The printing temperature is too low. The filament is not melted properly.

Solution(s):

i. Heat up the nozzle to approx. 250° C (480° F) and clean it using an acupuncture needle (Attention: danger of burns!).
ii. Reduce the feed rate in the slicing settings.
iii. Replace the nozzle completely (maybe the whole hot end) if the blockage is too heavy and / or the nozzle is excessively worn.
iv. Use an adequate printing temperature (see filament manufacturer).

7 Every Beginning is hard: Starting Issues

7.1 The 3D Printer or the Printing Process does not start

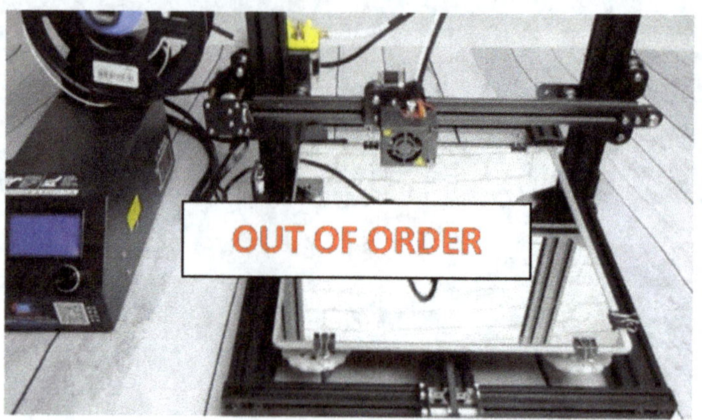

Figure 11: The 3D Printer does not work at all

Description:

The 3D Printer refuses to operate and does not start. The print head does not move, nor is filament extruded. However, heating of the nozzle/print bed may work.

Causes and Solutions:

Basic Information: *Make sure that the printer is plugged in and that all connectors and cables are intact. Also, check if there are any loose connections.*

i. Then check the file name of the print object first. If it contains an "umlaut" ("ä", "ö", ...) or other special characters (€, &, ...), some printers may refuse to print the file. Simply rename the object.

ii. If the printer does not reach the set temperature, it will not start either (see chapter 5.1 and 6.1).

iii. Check whether there is enough filament (an error message of the filament sensor can prevent starting).

iv. Use a different SD card or some other connection to the PC to exclude card failure/reading problems or general connection problems.

7.2 The 3D Printer does not extrude filament when printing starts

Figure 12: There is no filament coming out of the 3D Printer's nozzle

Description:

There is no filament coming out of the printing nozzle. However, the printer starts moving as usual after the heating process. This means that the print head is moving, and the fans are rotating.

Causes and Solutions:

Basic Information: Check whether there is enough filament and that it is correctly inserted and reaches the print head.

i.   Take a look at the stepper motor of the extruder, which is responsible for transporting the filament. Does it work? And is the filament sufficiently "gripped" and "transported"? **Also, note point "ii."**
ii. If the stepper motor cannot transport the filament, this may be due to mechanical issues in the area of the extruder or a clogged nozzle. Please read chapter 6.2 and 9.15.
iii. Check all other areas through which the filament passes or where there may be bottlenecks (e.g., Bowden tube and its connection points; hot end) or whether there is a knot in the filament.

7.3 Poor Print Bed Adhesion

Figure 13: The printed layers do not adhere

Description:

The object does not adhere to the print bed. This issue can occur while printing the first layer as well as later on – e.g., at a 50 % printing progress.

Causes and Solutions:

Basic Information: Clean the print bed using isopropyl alcohol or part cleaner.

i. The most probable cause of this issue is an incorrectly leveled print bed. Check the distance between the nozzle and the bed at several points by using a piece of paper (see chapter 4). If the paper can be moved easily, the distance is too large. Perform the leveling process one more time and reduce the distance slightly. If the distance is correct at the corners, but not in the middle or at the sides, the print bed may be deformed.

ii. Improper printing surface: If possible, use a mirror (plane and good adhesion) or a permanent printing plate, such as "BuildTak." For better adhesion, you can also put some tape or hairspray onto the print bed. But better avoid this if possible (results in a very high cleaning effort after each printing process).

iii. The printing bed is not heated, or the temperature is too low / too high. Use a heated bed in the range of 60° C (140° F) (see filament manufacturer's recommendations).

iv.   The printing temperature for the filament is too low. Use an adequate printing temperature (manufacturer's specifications) and check whether the 3D Printer maintains this temperature at a constant level. Reduce the power of the fans during the first print layer.

v.   Check the printing speed for the first few layers. Print at a lower speed (max. 30-50 mm/s).

vi.   Make sure that enough material is extruded during the first layer. Take a look at the nozzle and check that material flows continuously from the very beginning (also see chapter 8.1 & 9.1).

vii.   Use the print bed adhesion type: "Brim" or "Raft" as additional support to widen the contact area (especially for small areas). If necessary, also use the function "Skirt." These settings are made in the slicing software.

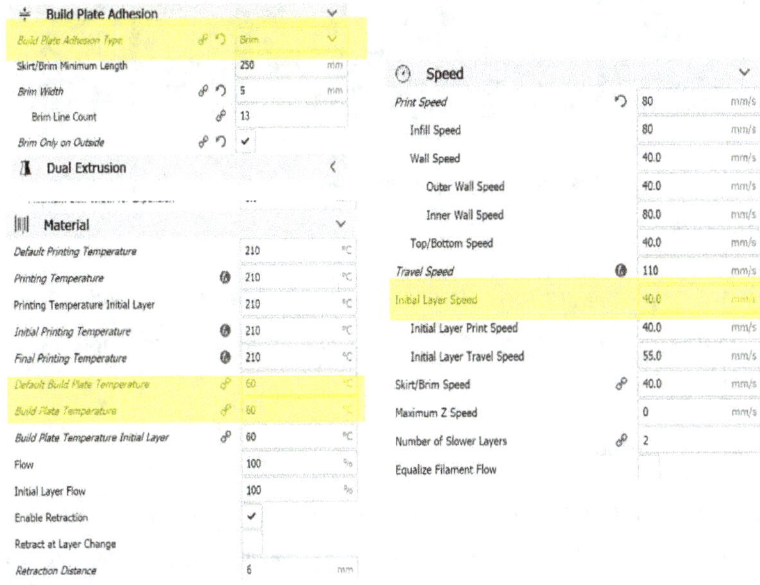

Figure 14: Sample settings for related slicing parameters

8 Filament-Issues

8.1 The Filament is not extruded in a Continuous Flow

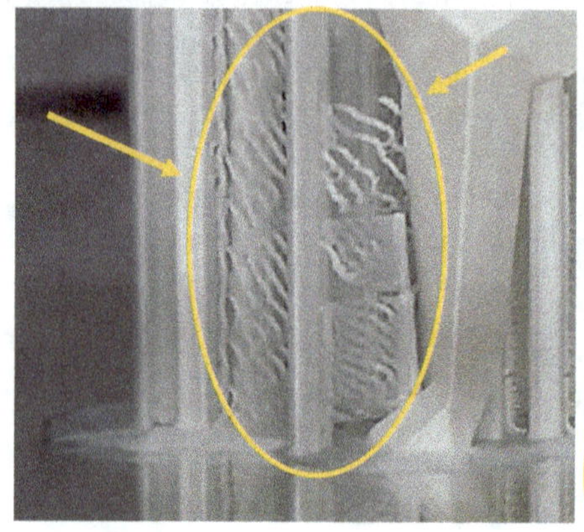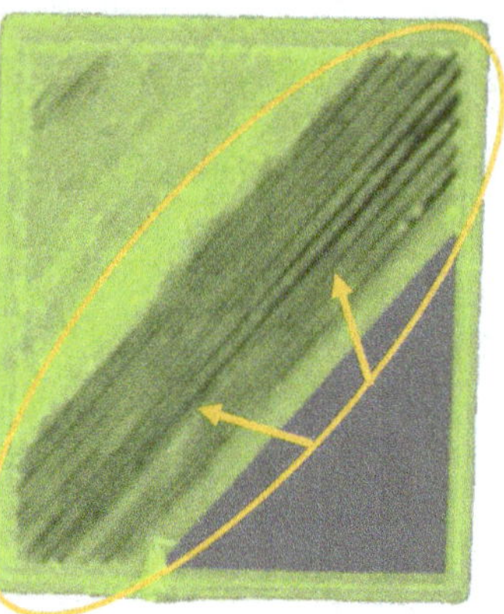

Figure 15: The filament is not extruded in a continuous flow

Description:

During the printing process, an insufficient amount of filament is extruded. There are gaps or a lack of material in the printed object. Take a look at chapter 9.1 as well.

Causes and Solutions:

Basic Information: Don't skimp on the filament. If the filament is very cheap or comes from a relatively unknown manufacturer, the filament quality may be the cause of the problem. In this context, the tolerance of the filament diameter is one of the most important factors. If the tolerance is high (≥ ± 0.05 mm), this may contribute to the shortage issues. (Recommendation: Use Filament from "Sunlu" or "Tianse")

i. Make sure that you are not dealing with a blocked nozzle (see chapters 6.2 & 9.15).

ii. Check that the filament supply is running smoothly, i.e., that the filament spool has sufficient space for movement (rotation) and that the filament can slide freely within the Bowden tube.

iii. Check the following slicing settings:
1) Set filament diameter: e.g., 1.75 mm.
2) Adequate printing temperature.
3) Material flow: 100% (increase if necessary) & decease retraction if necessary.

iv. Check whether the extruder grips the filament properly (usually with a small gear wheel). If necessary, increase the tension of the fixing device – if possible – or replace the extruder if the filament is not transported properly.

‖‖‖ Material			⌄
Default Printing Temperature		210	°C
Printing Temperature	🝳	210	°C
Printing Temperature Initial Layer		210	°C
Initial Printing Temperature	🝳	210	°C
Final Printing Temperature	🝳	210	°C
Default Build Plate Temperature	✑	60	°C
Build Plate Temperature	✑	60	°C
Build Plate Temperature Initial Layer	✑	60	°C
Flow		100	%
Initial Layer Flow		100	%
Enable Retraction		✓	
Retract at Layer Change			
Retraction Distance		6	mm

Figure 16: Sample settings for the parameters: "Flow" and "Retraction."

8.2 The Filament is brittle / breaks during Printing

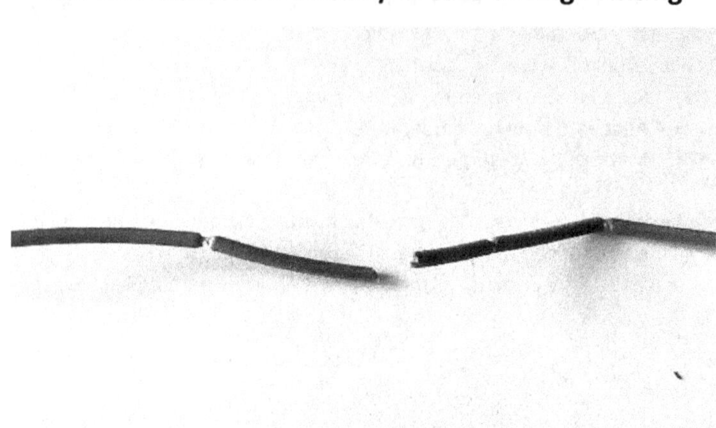

Figure 17: Brittle and broken filament

Description:

The filament is very brittle and breaks at slight bending or during the printing process.

Causes and Solutions:

Basic Information: *Store your filament in a dry (filament absorbs moisture), dark (UV radiation is very bad) and correctly tempered area (e.g., in a box with a silicate bag and a room with low humidity and temperature in the range of 20 - 23 ° C or 68 - 74 ° F).*

i.   The filament has been stored incorrectly or has been in use for a few years. Replace the filament if it is older than one year (when the package was opened).

ii.   Too much bending during filament transport: If the Bowden tube has been installed with strong changes of direction, the filament may be bent too much. Install the Bowden tube with as few bends as possible.

iii.   The position of the filament spool is not well selected. If the filament spool is placed at a very low level and far away from the extruder, the filament may also break. Choose a position at the same height or above the extruder and make sure that the filament transport is carried out very smoothly.

8.3 Filament Grinding / Stripping / Crushing

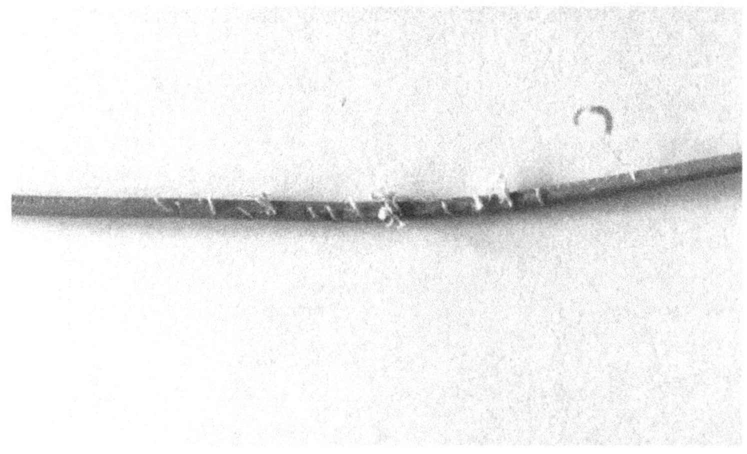

Figure 18: A piece of filament damaged by the extruder

Description:

The filament shows signs of material abrasion and / or looks pretty worn. Indentations are also visible.

Causes and Solutions:

Basic Information: *Use high-quality filament.*

i.   The extruder of an FDM 3D Printer usually fixes the filament by means of a small gear wheel. This often leaves only small visible indentations. However, if the extruder tries to push the filament excessively due to a blockage, the filament can be literally maltreated at this point. The reason for this can be a clogged nozzle or any other blockage (see also chapters 6.2, 7.2 and 9.15)

ii.   Check your slicing settings:

    1)   Too much and too fast retraction.

    2)   The printing speed is too high. If the printing speed is too high, the printing nozzle will not be able to cope with the amount of material supplied, and the pressure building up in the hot end may be too high. Reduce the printing speed.

    3)   Printing temperature too low: the filament cannot be sufficiently melted. Same situation as in 2).

iii.   Print platform is not leveled correctly. If the nozzle is too close to the print bed, this can also be a reason causing material jams in the nozzle or hot end.

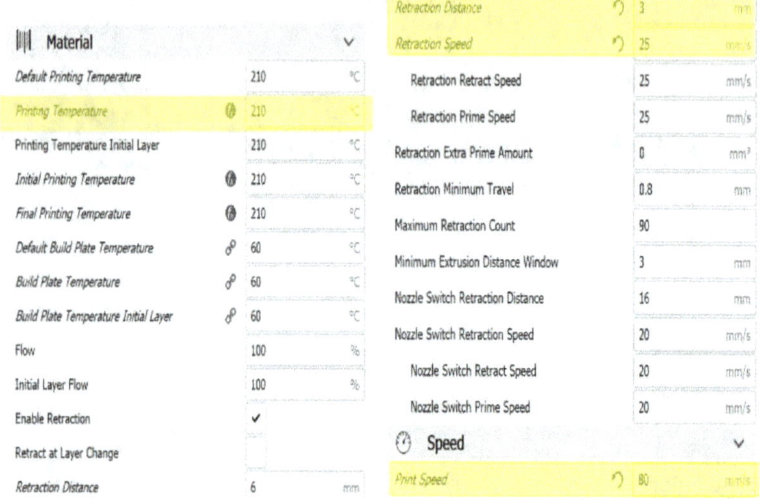

Figure 19: Sample settings for the parameters: "Print temperature," "Retraction," and "Print speed."

9 Particular error patterns

9.1 Under-Extrusion

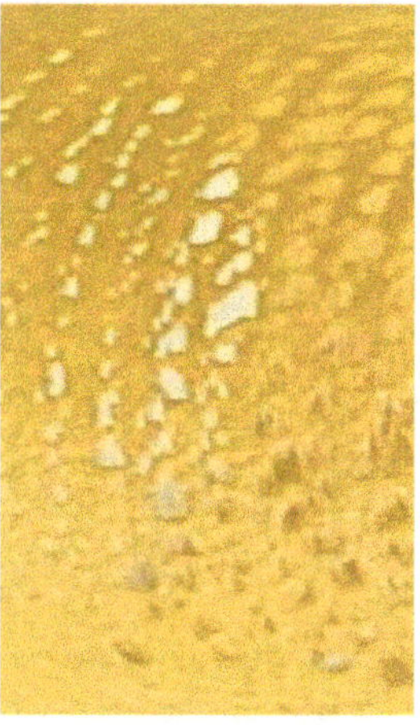

Figure 20: Under-extrusion occurring in the support structure (left)

Description:

An insufficient amount of material flows out of the nozzle. The printed structures are very thin, incomplete, and insufficiently formed (very similar to chapter 8.1).

Causes and Solutions:

Basic Information: *often very similar to the symptoms of a partially clogged nozzle.*

i.   Check all points in chapters 6.2, 8.1 & 9.15, as mechanical bottlenecks or a (partially) blocked nozzle are often possible causes.

ii.   Use the following recommendations in the slicing software:

    1) Choose a higher print temperature (download a "Temperature Tower" from thingiverse.com) and lower the print speed.

    2) Check the set filament diameter (e.g., 1.75 mm).

    3) Set the filament flow to 100% and increase it by a few percentage points if necessary.

    4) Reduce the retraction speed to a minimum and additionally reduce the retraction distance.

5) If the error pattern only occurs in the area of support structures or filling, change the following settings for these structures individually: Print speed (lower) & density (higher).

Default Printing Temperature		210	°C
Printing Temperature		210	°C
Printing Temperature Initial Layer		210	°C
Initial Printing Temperature		210	°C
Final Printing Temperature		210	°C
Default Build Plate Temperature		60	°C
Build Plate Temperature		60	°C
Build Plate Temperature Initial Layer		60	°C
Flow		100	%
Initial Layer Flow		100	%
Enable Retraction		✓	
Retract at Layer Change			
Retraction Distance		2	mm
Retraction Speed		25	mm/s
Retraction Retract Speed		25	mm/s

Figure 21: Settings for "Retraction" and "Print temperature."

9.2 Over-Extrusion

Figure 22: Over-Extrusion occurring on the sides of a cube

Description:

Too much material flows out of the nozzle. The printed structures are oversized and have an unclean surface.

Causes and Solutions:

Basic Information: *The opposite of under-extrusion.*

i. Try the following recommendations regarding the slicing settings:
   1) The printing temperature is too high (reduce it by 10-20° C (50-70° F); print a "Temperature Tower" (thingiverse.com) to find the optimal temperature for printing with the filament you have.
   2) Check the set filament diameter: e.g., 1.75 mm.
   3) Set the filament flow to 100% and reduce it by a few percentage points if necessary.
   4) Increase the retraction speed and distance.
ii. Calibrate the value of: "Steps per mm" of your extruder stepper motor (see manufacturer's instructions).

9.3 Curling

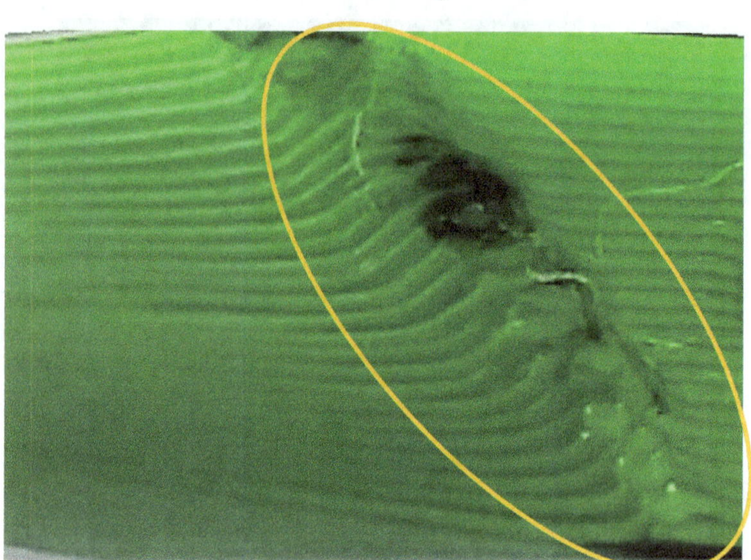

Figure 23: Curling at the corners

Description:

The edges, and especially the corners of the printed object, curl upwards.

Causes and Solutions:

Basic Information: *This error pattern is mainly caused by a too high printing temperature.*

i. Try the following recommendations regarding the slicing settings:
    1) The printing temperature is too high (reduce it by 10-20° C (50-70° F); print a "Temperature Tower" (thingiverse.com) to find the optimal temperature for printing with the filament you work with.
    2) Increase the fan speed or raise it to the maximum possible. Moreover, check whether the fans are operating properly.
ii. Check your room temperature and open the 3D Printer's enclosure

9.4 Stringing or Oozing

Figure 24: Classic Stringing Test

Description:

Web-like structures are formed in the area of the printed object. Reason: Plastic is still being extruded, although the nozzle is already moving to the next point.

Causes and Solutions:

Basic Information: *If you cannot get the error under control, you can remove these structures relatively easily instead.*

i. Try the following recommendations regarding the slicing settings:
    1) The printing temperature is too high (reduce it by 10-20° C (50-70° F).
    2) Increase the fan speed or raise it to the maximum possible. Also, check whether they are operating properly.
    3) Activate the function "retraction" and set it to reasonable values (retraction distance: 2-10 mm; retraction speed: 40-80 mm/s). This

function removes the pressure that builds up in the nozzle by retracting the filament while moving to the next printing area, and thus preventing an undesired outflow of material.

4) This printing error is also often caused because of too long traveling distances between several print objects. Place the print objects as close as possible to each other when slicing. It is also recommended to print only one object per job.

5) Additionally, activate the function "Coasting." The last movement of the print head is then carried out without any further material supply so that that excess material can be used at the end of a print movement.

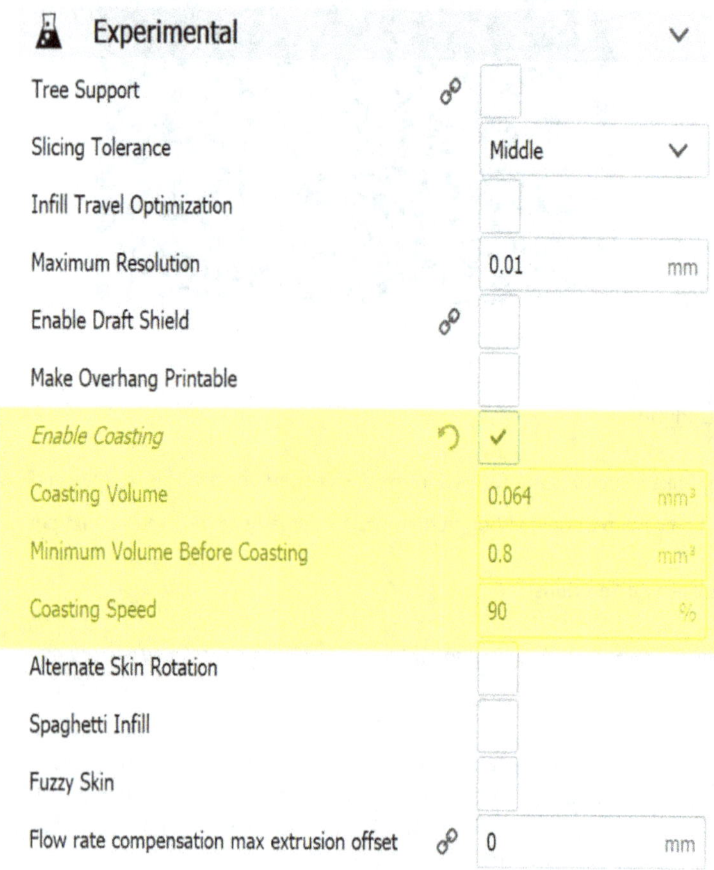

Figure 25: Activate "Coasting"

9.5 Blobs and Zits

Figure 26: Blobs & Zits on the outer wall of a 3D printed object

Description:

There are isolated and small material deposits in the form of droplets or "pimples" on the outer walls of the printed model.

Causes and Solutions:

Basic Information: If you cannot solve the problem, you can retouch these deposits by finishing with fine-grained sandpaper.

i.   Try the following recommendations regarding the slicing settings:

    1)   Check the retraction settings. The main cause of the problem is usually an incorrect value in this section. Recommended values for the retraction distance are: 2-10 mm; retraction speed: 40-80 mm/s. Avoid too many retractions by increasing the "Minimum distance for retractions" or decreasing the "Maximum number of retractions."

    2)   Set the Combing Mode to: "Within Filling." This will cause the print head to move as much as possible in areas that have already been printed (movements to the next print area will be within the filling). This reduces the number of retractions and segregates excess material within the filling.

3) Deactivate "z-hop when retracted." This function allows the platform to be lowered or raised briefly while retracting in order to create a distance between the nozzle and the object.

4) "Blobs" & "Zits" often occur due to the settings for the z-seam. The z-seam is a vertical structure that is visible on the outside of a print object. The starting point of a layer is where material often accumulates, resulting in a vertical "seam." Change the setting of the "z-seam position" from "random" to a defined value (x, y coordinate), which is located in a relatively invisible area of the object (e.g., back). Then the printer will not start the layer at a random position (which can lead to the "blobs" on the object walls) but will pull up the "seam" at a vertical position. Completely getting rid of the z-seam is often very difficult.

ii. Make sure to use dry and relatively new filament and follow the instructions for storing opened filament given at the beginning of the book.

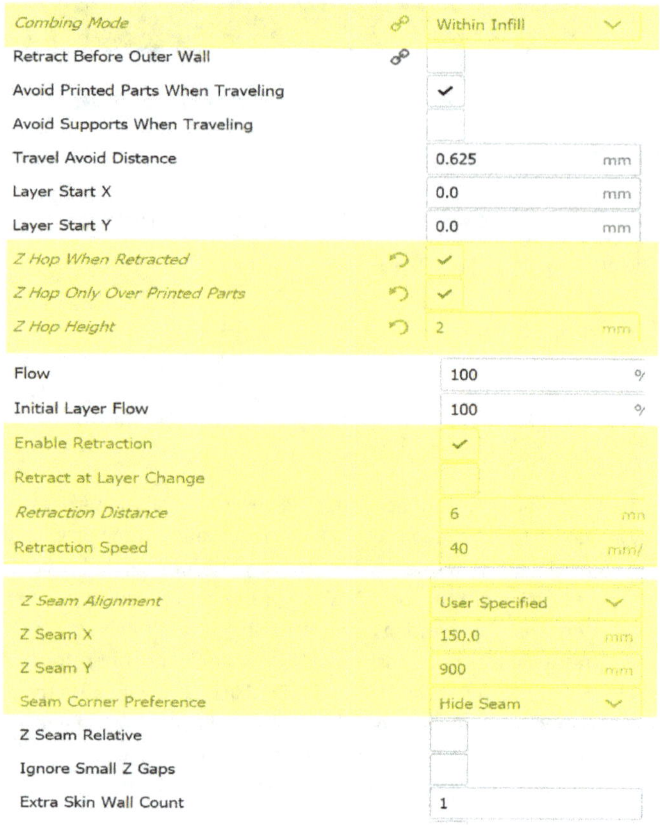

Combing Mode		Within Infill	
Retract Before Outer Wall			
Avoid Printed Parts When Traveling		✔	
Avoid Supports When Traveling			
Travel Avoid Distance		0.625	mm
Layer Start X		0.0	mm
Layer Start Y		0.0	mm
Z Hop When Retracted		✔	
Z Hop Only Over Printed Parts		✔	
Z Hop Height		2	mm
Flow		100	%
Initial Layer Flow		100	%
Enable Retraction		✔	
Retract at Layer Change			
Retraction Distance		6	mm
Retraction Speed		40	mm/
Z Seam Alignment		User Specified	
Z Seam X		150.0	mm
Z Seam Y		900	mm
Seam Corner Preference		Hide Seam	
Z Seam Relative			
Ignore Small Z Gaps			
Extra Skin Wall Count		1	

9.6 Pillowing

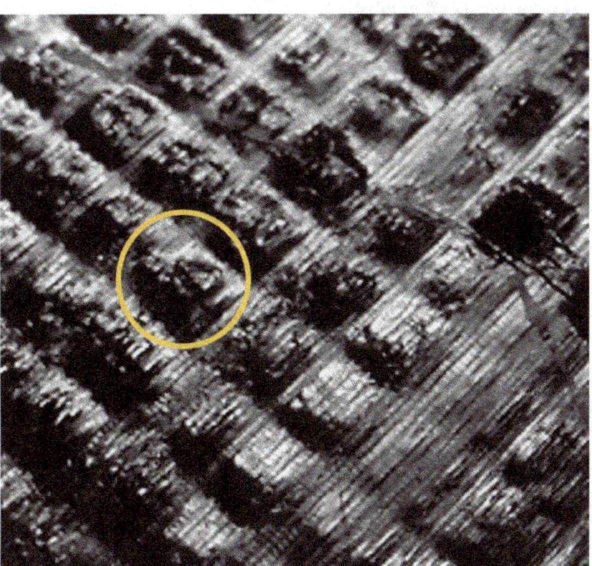

Figure 27: Small cushion-shaped dents or bumps on the printing surface

Description:

There are dents/bumps on the upper side of the print and/or the surface has holes.

Causes and Solutions:

Basic Information: *This is usually the result of an insufficient structure beneath the surface layer (i.e., a bad infill: see chapter 11.1) or too few surface layers.*

i. Try the following recommendations regarding the slicing settings:
    1) Use a higher infill percentage. A filling percentage of less than 15 % is usually not very useful. Also, try a different filling pattern and lower the printing speed of the filling.
    2) Reduce the print temperature (and increase the fan speed if necessary) and also reduce the overall print speed.
    3) Furthermore, check the suggested solutions of "Under-Extrusion" from chapter 9.1 and chapter 8.1.

4) Increase the number of top layers (e.g., three to seven layers will produce a good quality).

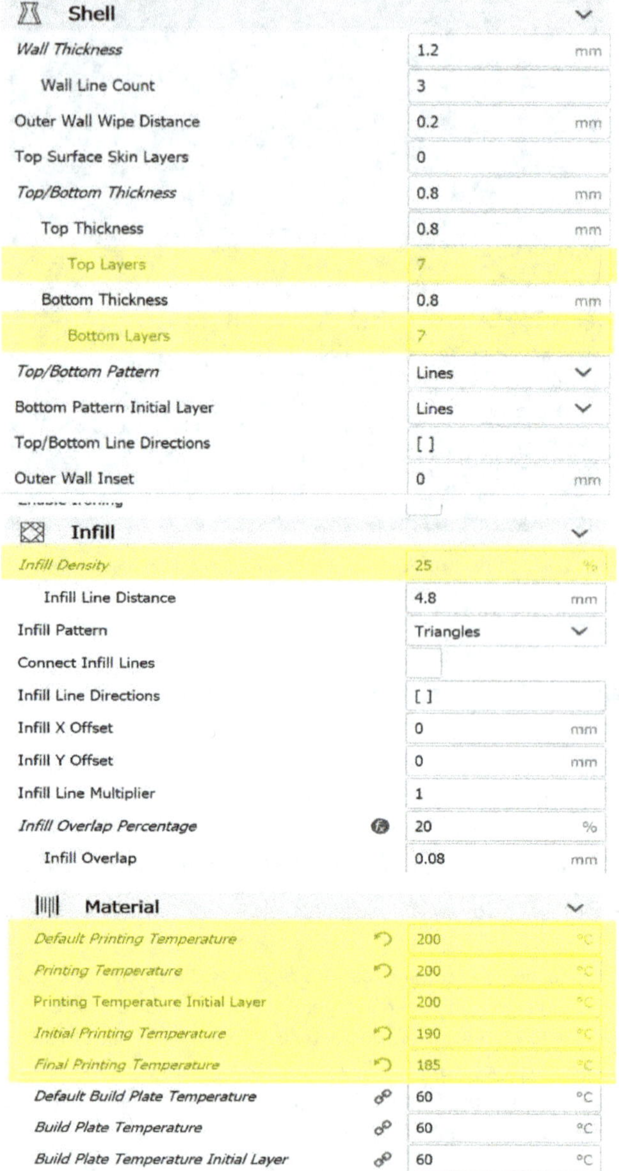

Figure 28: Sample settings for error-specific parameters

9.7 Vibrations & Ringing (Ghosting)

Figure 29: Vibrations & Ringing on the surface of a calibration cube

Description:

There are vibration-caused structures on the sidewalls of the printed object in the form of wavy contours, irregularities, and "shadowing."

Causes and Solutions:

Basic Information: *Mechanical causes can play a major role here. The frame of the printer and its stand should have as much stiffness as possible (see basic tips).*

i.　As already mentioned at the beginning: If the stiffness of the frame is too low because there are no adequate struts, vibrations can easily build up and be transmitted to the object to be printed.

ii.   Mount shock absorbers on the feet of the 3D Printer (Tip: Squash balls in combination with 3D-printed mounts).

iii.   The higher the mass of the hot end, the higher the reaction moments, and the higher the vibrations can be. Try to keep the mass as low as possible.

iv.   Also, check all ball bearings and drive belts for sufficient tightness.

v.   Check the suspension of the stepper motors. If the vibrations from the motors are insufficiently decoupled, they will be passed on to the 3D Printer frame, the print nozzle, and thus also to the object. Mount some dampening elements between the stepper motor and the 3D-printing frame.

vi.   Try the following recommendations regarding the slicing settings:

    1)   Causes in the slicing settings are to be found in the print speed and especially in the print acceleration. Sudden and rapid changes in movement are a problem. Reduce the overall print speed (e.g., 60-100 mm/s) and acceleration (max. 500 mm/s²).

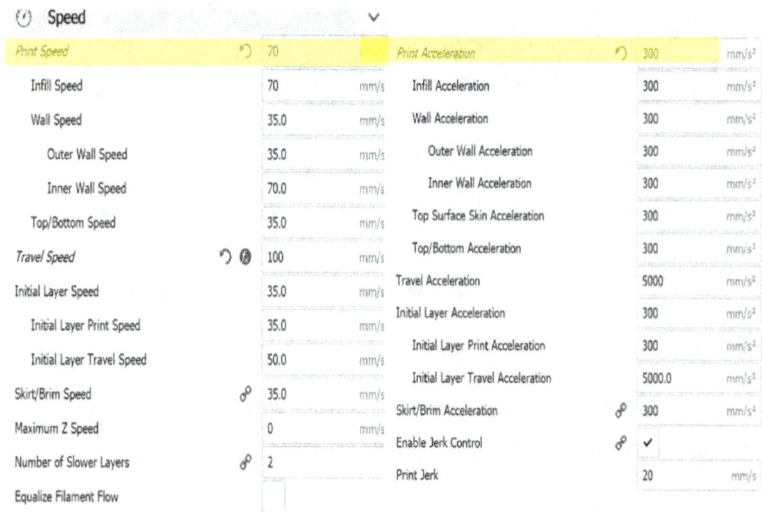

Speed			Acceleration		
Print Speed	70		Print Acceleration	300	mm/s²
Infill Speed	70	mm/s	Infill Acceleration	300	mm/s²
Wall Speed	35.0	mm/s	Wall Acceleration	300	mm/s²
Outer Wall Speed	35.0	mm/s	Outer Wall Acceleration	300	mm/s²
Inner Wall Speed	70.0	mm/s	Inner Wall Acceleration	300	mm/s²
Top/Bottom Speed	35.0	mm/s	Top Surface Skin Acceleration	300	mm/s²
Travel Speed	100	mm/s	Top/Bottom Acceleration	300	mm/s²
Initial Layer Speed	35.0	mm/s	Travel Acceleration	5000	mm/s²
Initial Layer Print Speed	35.0	mm/s	Initial Layer Acceleration	300	mm/s²
Initial Layer Travel Speed	50.0	mm/s	Initial Layer Print Acceleration	300	mm/s²
Skirt/Brim Speed	35.0	mm/s	Initial Layer Travel Acceleration	5000.0	mm/s²
Maximum Z Speed	0	mm/s	Skirt/Brim Acceleration	300	mm/s²
Number of Slower Layers	2		Enable Jerk Control	✓	
Equalize Filament Flow			Print Jerk	20	mm/s

Figure 30: Sample settings for the parameters: "print speed" and "print acceleration."

9.8 Warping

Figure 31: A corner of the print object detaches from the print bed (arrow)

Description:

Warping describes that one or more corners or edges of a print detach from the print bed and deform upwards.

Causes and Solutions:

Basic Information: *When the plastic – still warm from printing – gradually cools down, the material shrinks. This results in forces that are oriented in the opposite direction to the printing bed. If the adhesion forces are lower than these, the material will detach from the printing bed.*

i.   It is vital to use a heated bed with a printing bed temperature in the range of 60° C (140° F) (see instructions of filament manufacturer), especially for larger printing processes.

ii.   Mount an enclosure around the 3D Printer or close all doors of the enclosure to keep the heat inside.

iii.   Also, check the bed leveling (usually the distance between nozzle and print bed is a bit too high when warping occurs).

iv.   Try the following recommendations regarding the slicing settings:

    1)   Reduce the speed of the fans (maybe only for the first layer).

    2)   Use the bed adhesion type: "Brim" or "Raft."

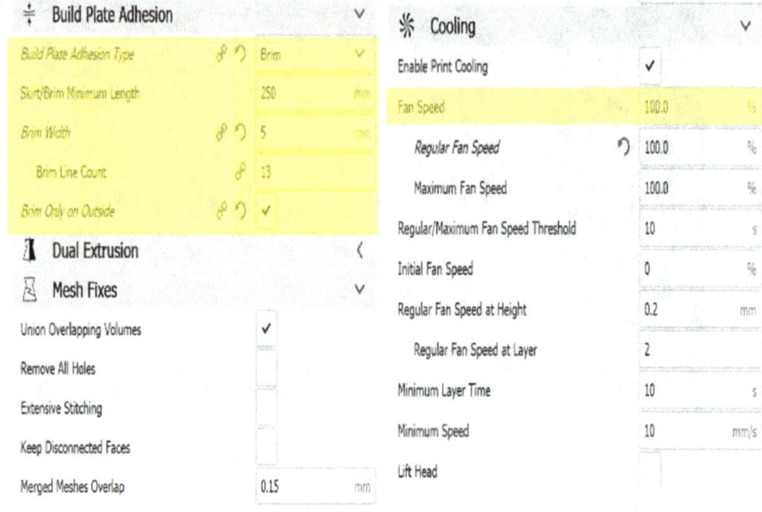

Figure 32: Settings for "Fan speed" and "Bed adhesion type."

9.9 Elephant Foot

Figure 33: An "elephant foot." All the weight is resting on the first few layers

Description:

The first layers of the print object are slightly wider than the following.

Causes and Solutions:

Basic Information: *This problem is usually due to the excessive influence of the printing bed heat. Depending on the filament, the optimum temperature can vary. The accumulating weight of the upper layers put forces on the lowest layer(s), and these – if they do not solidify sufficiently – give way.*

i. First, check the print bed leveling. This error may occur if the distance between the nozzle and print bed is too small.

ii.   Try the following recommendations regarding the slicing settings:
1) Reduce the speed of the fans during the first layer.
2) Lower the temperature of the print bed in the slicing settings. (see also: filament manufacturer's specifications).
3) Use the bed adhesion type: "Raft."

9.10 z-axis wobble

Figure 34: z-axis wobble; you can also see a z-seam here

Description:

With this kind of error, the walls of the print object are not printed smoothly. The layers are not correctly aligned on top of each other, but rather placed at a small distance in the x- & y-axis direction.

Causes and Solutions:

Basic Information: This pattern usually has mechanical causes.

i.   First, try to print with a lower layer thickness (e.g., 0.1 mm or 0.2 mm). Higher layer thicknesses generally lead to a bad print quality.

ii. If the problem remains, there may be a mechanical problem in the area of the z-axis. If the z-axis is out-of-round (slightly is enough) or is not guided well or with too high tolerances in the vertical direction, the z-axis will exert forces in the x & y-axis direction on the print head with every change in height (layer change). This causes the print head to shift slightly in the x- or y-axis direction, thus printing the layer with a small offset. Check the mechanical components for large deviations and replace them if necessary.

9.11 Layer Cracking / Separation / Splitting

Figure 35: Split layers in a 3D printed object

Description:

There are sharply defined gaps or layer separations in the print object due to insufficient connection (cohesion) between the layers.

Causes and Solutions:

Basic Information: *Sometimes also to be mistaken for "under-extrusion" (see chapter 9.1). The decisive factor for differentiation is whether the gaps are to be clearly defined or not. This error pattern is also comparable to "warping" (see chapter 9.8), with the difference of layer separations and deformations occurring while printing the layers of the object.*

i. Try the following recommendations regarding the slicing settings:

1) Since the filament must be melted sufficiently to adhere to the previous layer, it is advisable to increase the printing temperature a little bit (the printing of a "temperature tower" is also very helpful in this context).
2) Use a lower layer thickness (e.g., 0.1 mm or 0.15 mm).
3) Reduce the speed of the fans if necessary.

9.12 Layer Shifting

Figure 36: Shifted layers (3DBenchy of "creativetools" from thingiverse.com)

Description:

One or more layers are completely displaced or slightly offset from each other.

Causes and Solutions:

Basic Information: *If you have not created the "gcode" yourself, check if the error is due to the file.*

i. First of all, exclude mechanical causes: It may happen that the drive belts loosen or slip after some time. Check whether the belts are sufficiently tensioned and whether they reliably transmit the movements of the stepper motors. Likewise, it is possible that the belt pulley (connection between stepper motor and drive belt) is loose or slips from time to time.
ii. If there aren't any mechanical issues, use the following recommendation for the slicing:
iii. If you are printing at a very high speed, the stepper motors may not be able to handle it (often clicking noises can be heard). Reduce the overall printing speed by 30-50%.

9.13 Missing Layers

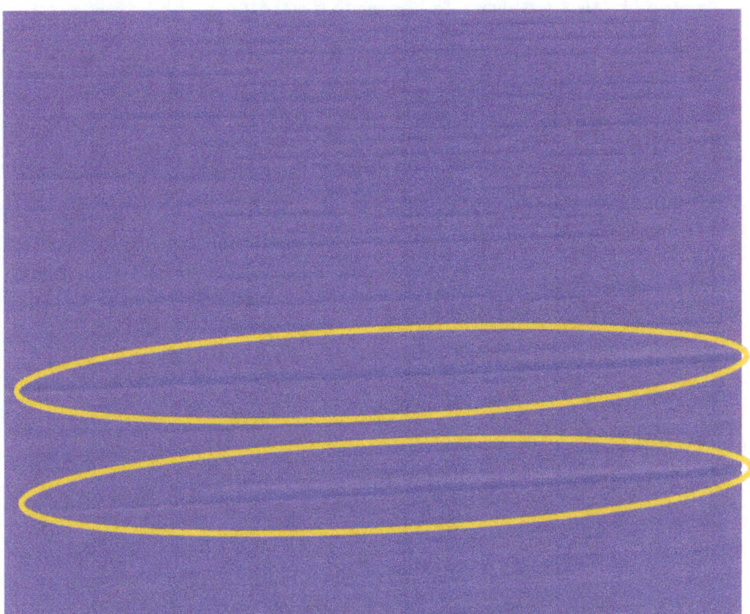

Figure 37: Some layers are missing in the printed object

Description:

One or more layers are missing in the printed object. This may occur several times in the print at different heights.

Causes and Solutions:

Basic Information: *Easily mistaken for "Under-Extrusion" (chapter 9.1) or "Layer Splitting" (chapter 9.11). However, missing layers are usually very distinct and not visible throughout the print, and especially do not show any upwards deformation.*

i.   This problem is usually not a missing layer itself, but rather a too-high step in the z-direction when changing layers. On the one hand, this is possible due to a faulty z-axis; on the other hand, it is caused by irregularities regarding the stepper motor. Check the z-axis, especially if it is bent or otherwise damaged. Also, check whether the z-axis can move freely.

ii.   In the slicing settings, you can usually set the z-axis steps. Reduce this value a little bit. Also deactivate: "z-hop when retracted"

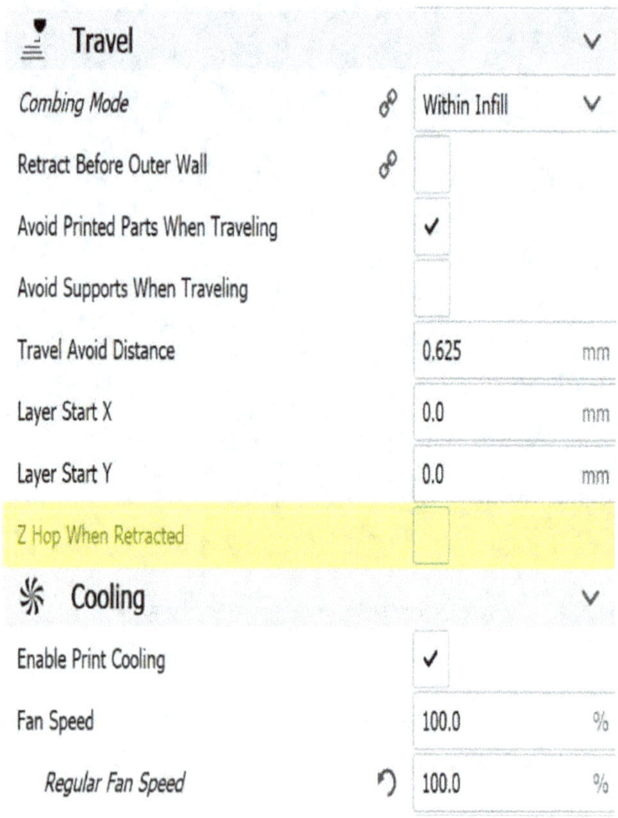

Figure 38: Deactivating "z-hop when retracted."

9.14 Walls are not smooth

Figure 39: Wavy and rough side walls

Description:

The side walls of the printed object are very rough; the individual layers are visible as lines.

Causes and Solutions:

Basic Information: Print with a low layer thickness (0.1 - 0.2 mm); this could already solve your problem.

i.   Check whether you are actually dealing with "Ringing & Vibrations" (see chapter 9.7).
ii. Furthermore, check if the problem is due to "under-extrusion" (chapter 9.1; see also 8.1).
iii. Take a look at the temperature on the 3D Printer's display. If the temperature of the printing nozzle fluctuates by +/- 10 degrees while printing, this may cause the issue. Also, make sure to avoid "on" and "off" switching of the fans.

If it is not due to the fans: contact the manufacturer of your printer and let them replace the PID controller if necessary.

iv. Check all mechanical components, especially the z-axis, for damage (also take a look at the solutions for "z-axis wobble": chapter 9.10).

v. Try to print at a reduced speed.

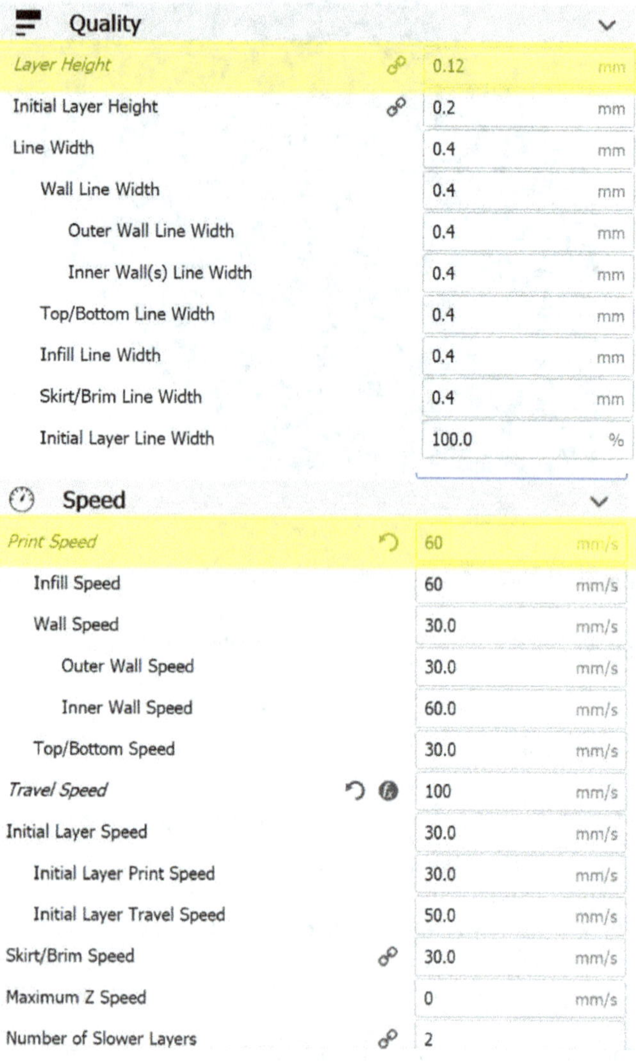

Figure 40: "Print speed "

9.15 Clogged Nozzle

Figure 41: There is no material coming out of the nozzle

Description:

There is no or very little material coming out of the nozzle. Clicking noises made by the extruder.

Causes and Solutions:

Basic Information: Make sure that the print bed is leveled. If the nozzle is too close to the print bed, this can clog the nozzle by increasing the pressure inside the nozzle.

Cause(s):

  i.   Accumulated residue in the nozzle.
  ii.   Feed rate too high, thus resulting in a material jam.
  iii. Mouth of the nozzle is blocked (wear/deformation due to mechanical damage).
  iv.   Printing temperature is too low: Filament is not getting melted in a proper way.

Solution(s):

i. Heat up the nozzle to approx. 250 ° C (480° F) (for PLA; with ABS or other materials even higher) and clean it with an acupuncture needle (Attention: danger of burns!). After heating up, you can also try to operate the extruder motor manually by the printer's menu and thus press material out of the nozzle.

ii. Reduce the feed rate in the slicing settings.

iii. Replace the nozzle completely (maybe also the hot end) if the blockage is too heavy and/or the nozzle is too worn.

iv. Use an adequate printing temperature (filament manufacturer).

9.16 Scars on the Print (Top / Bottom)

Figure 42: Traces of the nozzle appear on the print surface

Description:

There are visible lines or scratches on the top or bottom of the printed object.

Causes and Solutions:

Basic Information: The movements of the nozzle usually create these lines. When the hot nozzle moves over a printed layer that has already cooled down, the nozzle melts the plastic again and leaves a trace. On the other hand, such a trace can also be caused by superfluous secretions from the nozzle (see chapter 9.4).

i.   Try the following recommendations regarding the slicing settings:

    1) Activate the function "retraction" and set it to reasonable values (retraction distance: 2-10 mm; retraction speed: 40-80 mm/s). This function removes the pressure that builds up in the nozzle by retracting the filament during movements to the following areas and thereby preventing the unwanted material flow.

    2) Also, activate "Coasting." This allows the last movement of the print head to be carried out without any further material feeding so that the excess material can be used to print the last details.

    3) Activate the function "z-hop when retracted." This function allows the platform to be lowered or raised briefly in order to create a distance between the nozzle and the object.

    4) Use "Ironing." With this function, the nozzle moves over the top layer without any material flow (or with a small material flow; this can be set) and smooths it. In general, this produces a very high layer of quality.

However, "Ironing" only makes sense if the top layer is positioned relatively parallel to the print bed.

5) Use "Combing" and adjust the settings to "Not in Skin."

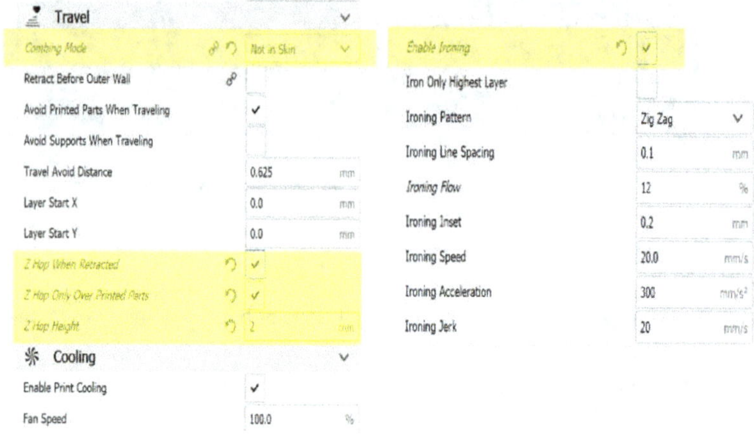

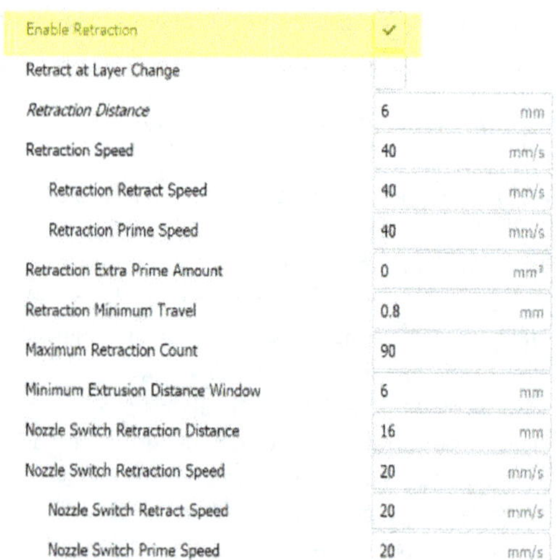

Figure 43: Specific slicing settings

9.17 3D Printer knocks over Prints

Description:

The print head knocks over the print. This usually happens at a more advanced print height.

Causes and Solutions:

Basic Information: *If the object has only a very limited contact to the print bed, and/or the center of gravity is relatively high (volume increases significantly with increasing height), the nozzle can knock over the object relatively easily during the printing process (for example by shaking movements as occurring in the case of smoothing/ironing). Especially thin and long elements that are predestined for this. It is sometimes impossible to print a long, thin cylindrical element in a vertical orientation.*

i. Check the leveling of the print bed (nozzle too close to the bed
ii. Try the following recommendations regarding the slicing settings:
   1) Activate "z-hop when retracted." This function allows the platform to be lowered or the nozzle to be raised briefly to create a distance between the nozzle and the object.
   2) Select a printing bed adhesion type (especially "Brim" and "Raft" are recommended here) to increase the contact area.
   3) Also, increase the number of Brim lines if necessary.
   4) Reduce the printing speed.

9.18 Poor Bridging of Gaps

Figure 44: Poor bridging of a somewhat longer distance

Description:

An FDM 3D Printer cannot print into the air. However, it usually manages to bridge smaller distances between two points without support material. If the quality of this bridging is very poor or does not work at all, read on here.

Causes and Solutions:

Basic Information: For longer distances, support material should always be used. Even if you cannot achieve success by following the steps below, you can still use support material to cope.

i.   Check the current bridging capability of your 3D Printer and slicing settings by loading and printing some simple "Bridging Test" object (thingiverse.com).

ii.   Try the following recommendations regarding the slicing settings:
1) Increase the relative fan speed to 100%.
2) Reduce the feed rate (sometimes an increased feed rate helps; try both).
3) Reduce the printing temperature.
4) Reduce the printing speed (sometimes faster is better; try both).

9.19 The Object cannot be removed from the Print Bed

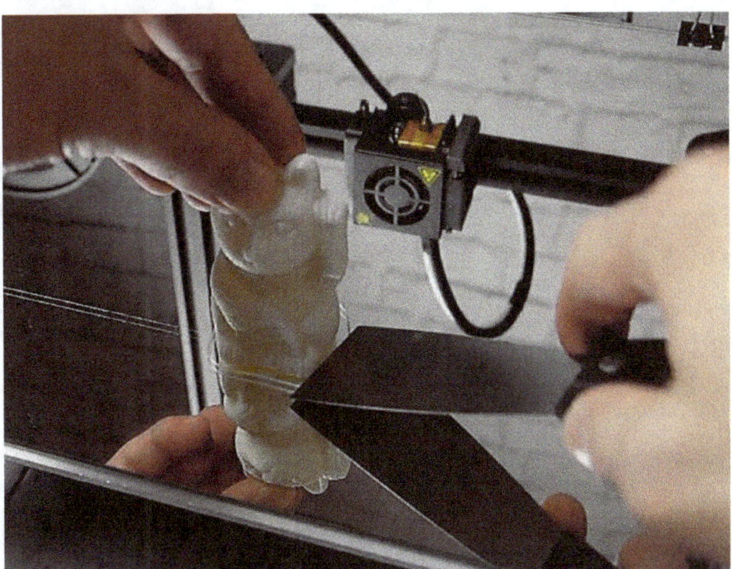

Figure 45: The printed object can only be removed by applying high force

Description:

When the printing process is done, the object cannot be removed from the print bed, or only with a great effort or by damaging the print (caution: risk of injury!).

Causes and Solutions:

Basic Information: Anyone who has already dealt with a poor print bed adhesion would probably consider this a luxury problem. However, it can be very dangerous for the 3D Printer, the printed object, and for yourself if the printed object doesn't want to come loose from the bed at all.

i. First, allow the heat bed to cool down completely; often, this resolves the issue by itself. When the plastic cools down, it shrinks a little and thus usually detaches from the bed.
ii. If not, the distance between the nozzle and the print bed may be too small. The importance of a well-leveled bed can be emphasized over and over again.

57

iii.   If you use additional aids to increase adhesion (tape, glue stick, hairspray, …), leave them out.

iv.   Try the following recommendations regarding the slicing settings:

    1)   Do not use an additional plate adhesion type (slicing settings; "Skirt" is fine).

    2)   Reduce the temperature of the printing bed by a few degrees.

    3)   Reduce the feed rate for the first print layer (or the general feed rate) and optimize all other settings for the first print layer (e.g., a higher layer thickness for the first layer).

v.   If none of these steps work, try cold spray.

‖‖ Material		
Default Printing Temperature	210	°C
Printing Temperature	210	°C
Printing Temperature Initial Layer	210	°C
Initial Printing Temperature	210	°C
Final Printing Temperature	210	°C
Default Build Plate Temperature	60	°C
Build Plate Temperature	60	°C
Build Plate Temperature Initial Layer	60	°C
Flow	100	%
Initial Layer Flow	100	%
Enable Retraction	✓	
Retract at Layer Change		
Retraction Distance	6	mm

Figure 46: Problem specific settings for "feed rate" and "printing temperature."

9.20 Gaps between the Filling and the Outer Walls

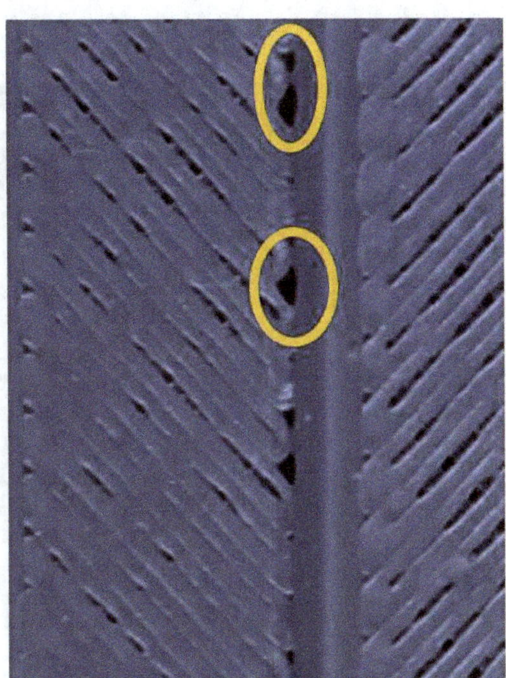

Figure 47: Gaps in the print surface

Description:

There are some visible gaps between the walls and the filling.

Causes and Solutions:

Basic Information: *Do not use too many different slicing settings for the walls and filling (especially for the specific speeds).*

i. Try the following recommendations regarding the slicing settings:

1) Reduce the print speed (filling & walls).
2) Increase the overlap percentage between inner walls and filling.
3) Try a different fill pattern.

9.21 The 3D Printer stops During Printing or at a Specific Height

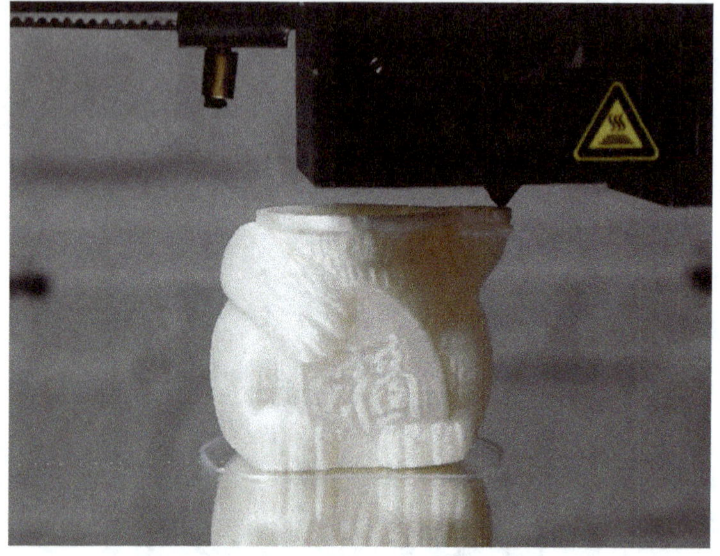

Figure 48: The 3D Printer suddenly stops printing

Description:

The printer stops in the middle of a printing process or repeatedly at a defined height and leaves the print job incomplete.

Basic Information: *Check the power supply.*

Causes and Solutions:

i. Check all mechanical components of the printer: especially the stepper motors, the connection couplings, the drive belts, the z-axis, and the extruder for proper function.

ii. There may also be a blockage in filament supply. Unfortunately, it can happen that the filament gets tangled or knotted on the spool. However, in this case, the printer would probably just keep moving without extruding the filament (and will possibly use the chance almost to destroy itself).

iii.   A defect of the filament sensor might stop the printing (if there is any sensor).

iv.   The nozzle can also be blocked (the movements of the print head would continue).

v.   There is a general overheating of 3D Printer components, e.g., the stepper motors. A thermo-switch prevents thermally induced damage and stops the printer for safety reasons.

vi.   If the printing process always stops at a certain height, this may be either due to defective z-axis or to incorrect settings in the slicing software (installation space set incorrectly; slicing process failed) or due to restrains (the wiring is too short; being restrained).

9.22 The z-seam is clearly visible

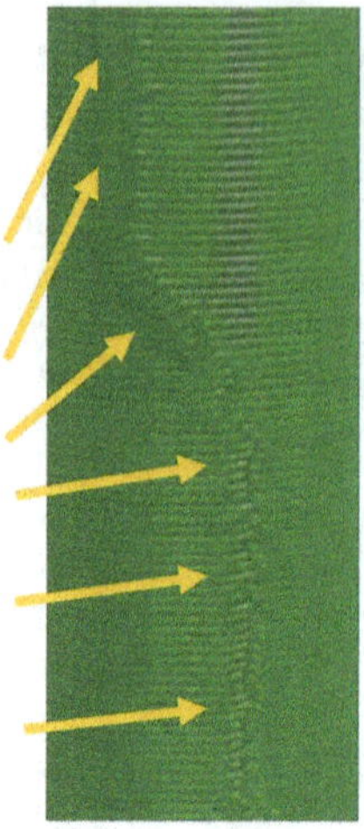

Figure 49: A "z-seam" has formed in vertical direction

Description:

The z-seam is prominent on the outer wall. The z-seam describes a vertical structure that is visible on the outside of a print object. It is formed because the 3D Printer must start at some point at the beginning of each layer.

Basic Information: It is very difficult to get rid of the z-seam totally. However, it is often possible to position it in inconspicuous areas, thus concealing it relatively well.

Causes and Solutions:

i. At the starting point of a layer, there is often an accumulation of material, resulting in a vertical "seam." Change the setting of the "z-seam position" from "random" to a defined value (x, y-coordinate) within a relatively invisible area of the object. Then the printer will not start each layer at a random position (which can lead to the "blobs" on the walls) but will pull up the "seam" at a vertical position. Avoiding such a seam completely is often very difficult.

10 Geometrical Problems

10.1 Dimensional Deviations from Print to CAD Model

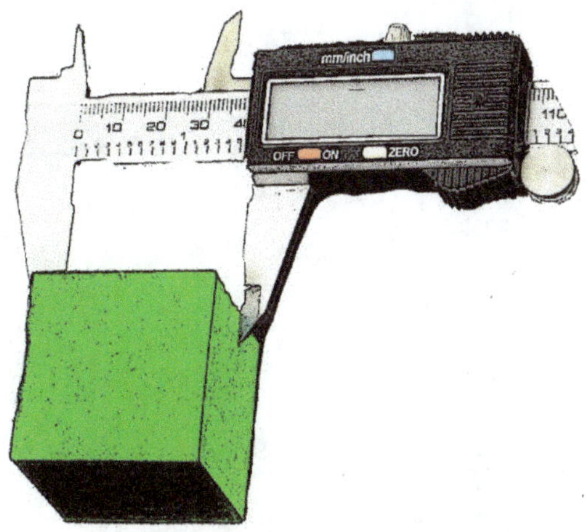

Figure 50: Measuring the component using a caliper

Description:

The printed object deviates (significantly) from the set dimensions.

Basic Information: Minor deviations are usually unavoidable. This is usually not so critical for the assembly of the parts if these deviations are constant throughout all printing jobs because the absolute dimensions of the elements are not exact, but the relative tolerances of the printed elements are kept and thus will fit each other. This does become critical; however, if the 3D printed objects are to be installed elsewhere or the dimensional deviations differ from print to print.

Causes and Solutions:

i. Check whether you have to deal with "under- or over-extrusion" (see chapters 9.1 & 9.2) (try the recommended steps if you are not sure).
ii. If only the first layers of the object are printed oversized, check chapter 9.9 ("Elephant foot").

iii.   If the problem persists, you can also set higher or lower tolerances during the design of the model, or simply scale the individual objects in the slicing software by 1-5 % larger or smaller.

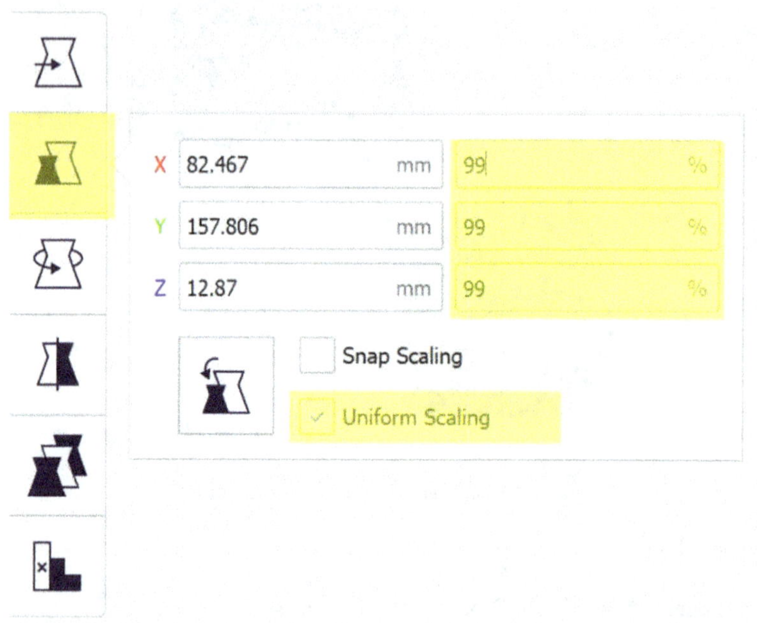

Figure 51: Scale print model by 1-2 % smaller or larger (uniform scale)

10.2 Circles are printed oval

Description:

Circles are not printed round, but oval.

Basic Information: *Minor deviations are usually unavoidable. Sometimes a different placement of the object can already be successful. The axis of a hole or circle should – if possible – be positioned vertically, so that the influence of gravity isn't a factor.*

Causes and Solutions:

i.   Check for mechanical issues: Are the drive belts sufficiently tensioned? Do the drive belts transmit the movements of the stepper motors? Is the z-axis ok?

ii.   Reduce the printing speed.

iii.   Try a different positioning (especially: rotations) of the object when slicing.

10.3 Thin Elements are not printed

Description:

Very thin or small elements are not printed.

__Basic Information:__ If you print elements (e.g., 0.3 mm) that are thinner than the nozzle diameter (e.g., 0.4 mm nozzle), they may simply not be printed.

Causes and Solutions:

i.   If possible, try not to use or print structures thinner than the nozzle diameter of your 3D Printer. Alternatively, use a smaller nozzle diameter.
ii.   Moreover, activate the setting: "Print thin structures" or a comparable slicing setting.
iii.   Check if you can change the position of the object (when slicing) or if scaling is possible.

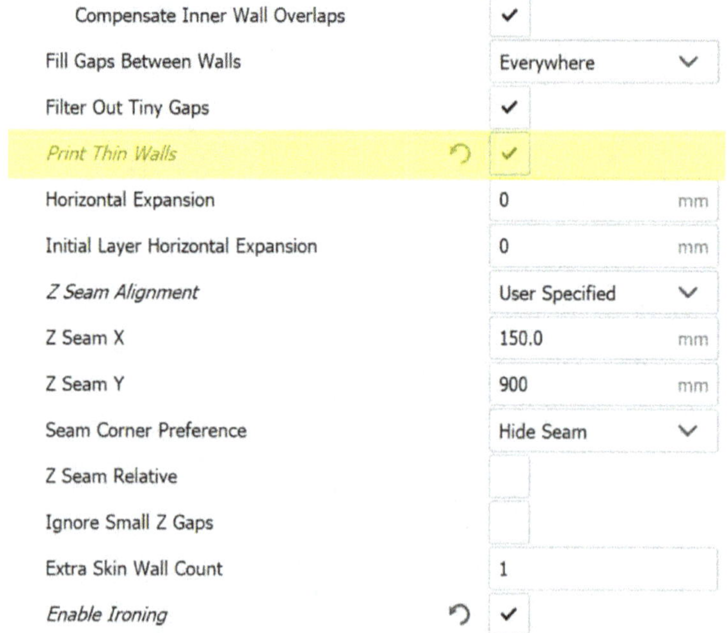

Figure 52: Activate "Print Thin Walls"

11 Problems regarding the Infill

11.1 Generally poor Infill

Figure 53: Poorly printed infill pattern

Description:

The filling of the object is printed qualitatively/quantitatively very badly, has gaps or other defects.

Basic Information: *If the component is not exposed to any loads, and the top layer of the object is printed in good quality, this problem is negligible.*

Causes and Solutions:

i.   Try the following recommendations regarding the slicing settings:
   1) Reduce the print speed, especially the specific infill print speed (if possible).
   2) Check all steps from chapter 9.1 ("Under-Extrusion").
   3) Increase the filling density (to 20 - 30%).
   4) Increase the layer thickness of the filling (if possible).
   5) Try a different fill pattern.

11.2 The Infill shines through the Object and is visible from Outside

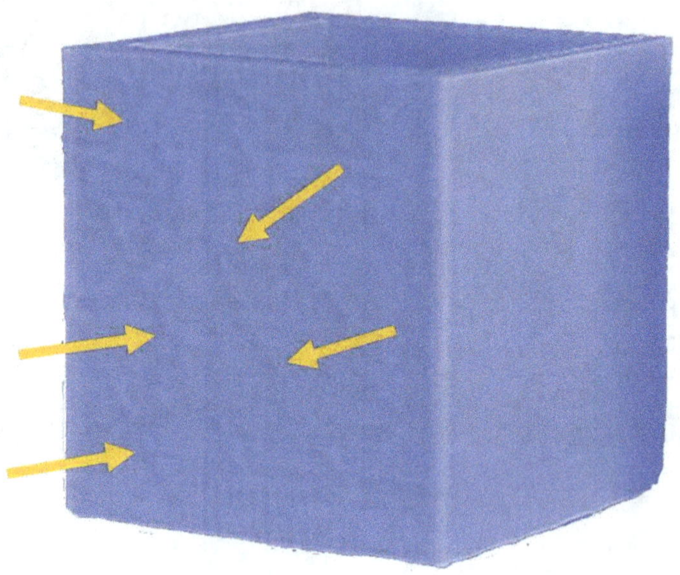

Figure 54: Take a close look, and you will notice the filling pattern in the front area

Description:

The filling pattern of the print object is visible from outside.

Basic Information: *An appropriate number of outer walls will increase the printing time but should solve the problem to a satisfying level.*

Causes and Solutions:

i.   Try the following recommendations regarding the slicing settings:
1) Increase the number of outer walls (2-3 are optimal; results in a 0.8 - 1.2 mm wall thickness using a 0.4 mm nozzle and line width).
2) The function: "Print outer walls before inner walls" can also be a remedy.

12 Problems regarding the Support Structure

12.1 The Support Structure is (nearly) impossible to remove

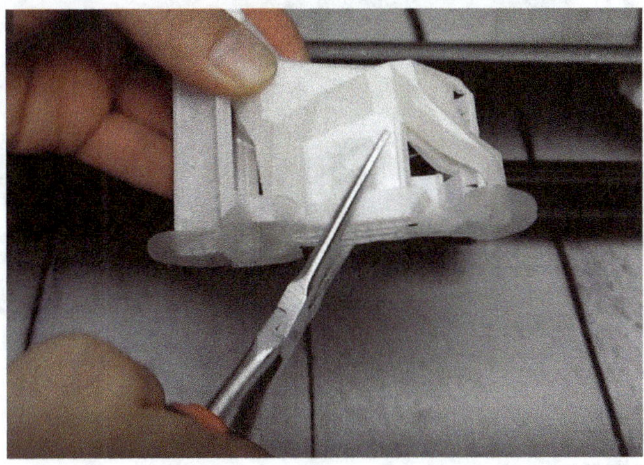

Figure 55: The support structure can only be removed by applying high force

Description:

The support structure can only be removed by applying a high amount of energy or by damaging the printed object.

Basic Information: Always try to use as little support structure as possible. In general, you can print overhangs of up to 60° – with some printers only 45° – without support structures. A well-chosen positioning of the object on the printing bed creates a further potential for reducing the amount (rotations are often helpful).

Causes and Solutions:

i.   Try the following recommendations regarding the slicing settings:
1) Reduce the density of the support structure (e.g., to 10 - 15%).
2) Increase the distance between the support structure and the model (x, y, and z distances; values in the range of 0.1 - 1 mm are recommended).
3) Use a different pattern for the support structure.

12.2 Poor Surface Quality in the Area of Support Structures

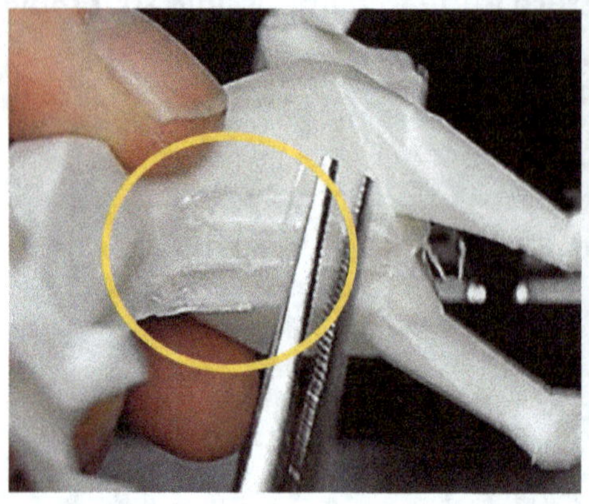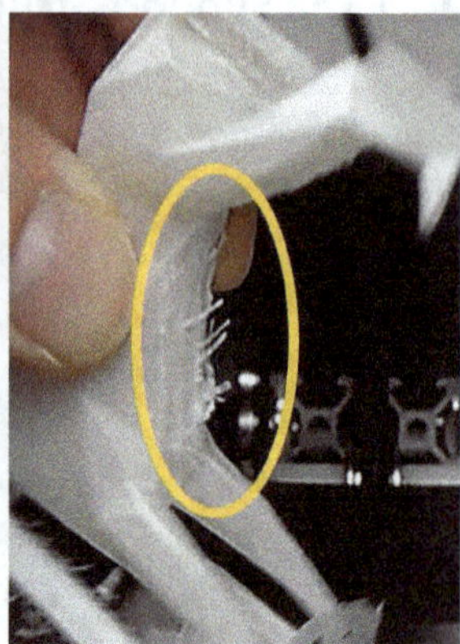

Figure 56: Rough and difficult to remove support structure

Description:

The surface quality of the printed object is very poor in the area of supports.

Basic Information: *This problem cannot usually be completely eliminated when using a support structure. If you have a dual extruder (i.e., a 3D Printer with two nozzles), you would be able to use water-soluble material (e.g., PVA) for printing the supports and then simply remove it after the printing process by washing it. This will typically result in a much higher quality.*

Causes and Solutions:

i. Try the following recommendations regarding the slicing settings:
    1) Reduce the overall layer thickness (0.1 mm typically produces very high-quality surfaces).
    2) Increase the density of the support structure (e.g., to 20 - 40%).
    3) Increase or decrease the distance between the support structure and the model (x, y, and z distances; values in the range of 0.1 - 1 mm are recommended).

12.3 Overhangs are not supported / Support Structure is of very poor Quality or falls away during printing

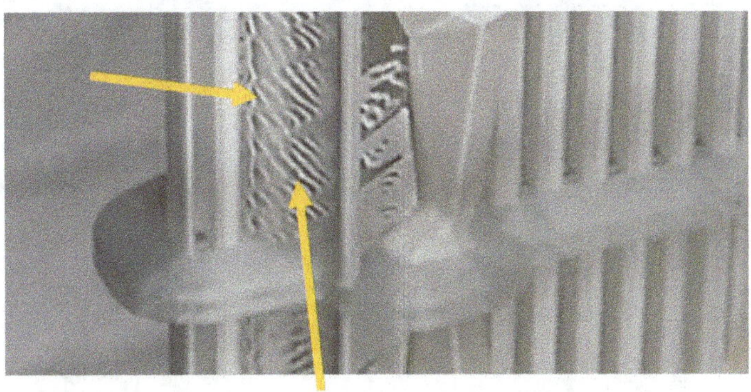

Figure 57: The support structure is printed in a very poor quality (left area)

Description:

The support structure already falls away during printing or does not support overhangs at all.

Basic Information: Contrary issue to section 12.1.

Causes and Solutions:

i. Try the following recommendations regarding the slicing settings:
    1) Increase the density of the support structure (e.g., to 20 - 30%).
    2) Reduce the distance between the support structure and the model (x, y, and z distances).
    3) Use a different pattern for the support structure.
    4) Use the bed adhesion type: "Brim" to increase the contact of the support structure to the print bed.
    5) Use the function: "Connect support structure" to link the individual structural elements.
    6) Reduce the print speed of the support structure.
    7) If overhangs are not supported, check the settings for the support angle of the structures. Reduce it to approx. 45°. If possible, with your slicing software, also set further support structures in the area of the overhangs.
    8) Use the special functions "Use Towers" or "Supporting roof."

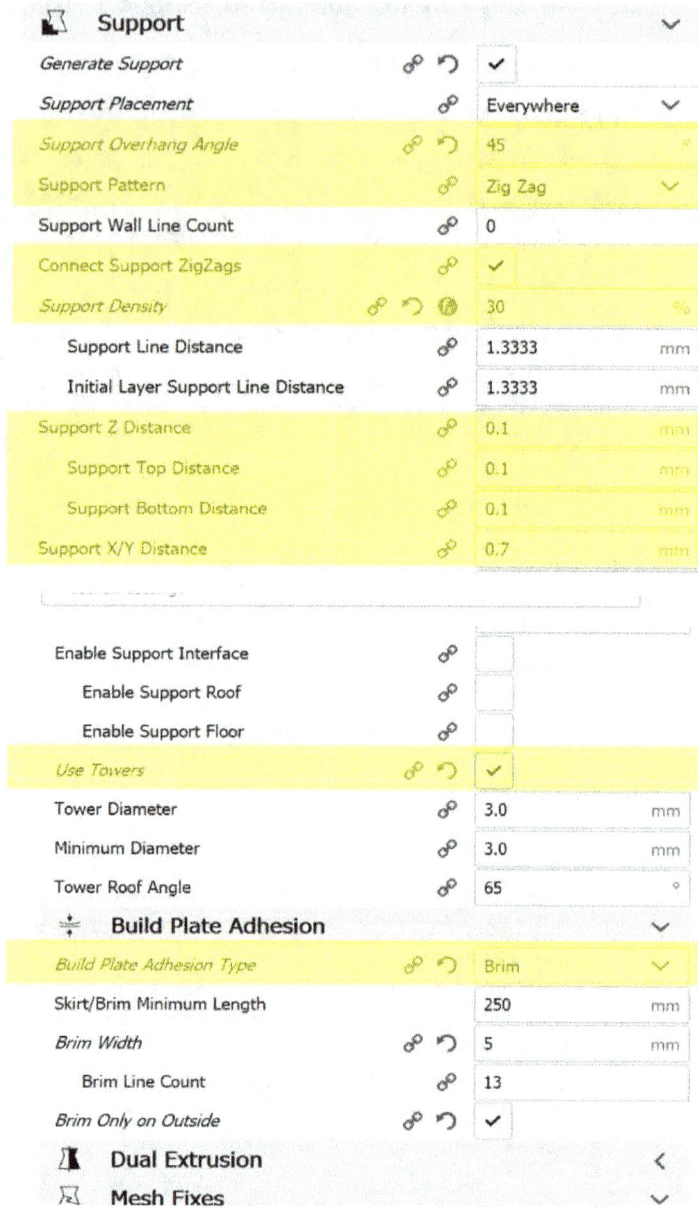

Figure 58: Specific settings for the support structure

13 The Quintessence – Master 3D Printing

Which settings have the greatest influence on print quality? If you're striving to push your 3D Printer to its peak regarding quality, you should take a closer look at the following settings and do some trial runs with different parameters:

Layer Thickness:

The layer thickness is probably one of the most crucial parameters when it comes to the print quality of the outer shell of an object. It is best to print at 0.1 mm when you have enough time. An insider tip also says that with a 0.4 mm nozzle, you should choose the thickness of the layer in 0.04 mm steps. I.e., 0.12 mm, 0.16 mm, 0.20 mm. Of course, the printing time can increase considerably with a lower layer thickness, which can be a "no-go" criterion for a 0.1 mm layer for larger objects.

Print Speed:

The overall print speed of a 3D Printer, as well as the specific speeds for outer walls, inner walls, infill, and print head movements, can have a massive impact on quality. A recommended print speed ranges from 60 - 100 mm/s for the overall speed and 30 - 50 mm/s for printing the walls and first layers. Furthermore, make sure that the respective parameters do not differ too much from each other.

Number of outer Walls:

If possible, print at least two outer walls per object, preferably even three or more. On the one hand, this prevents the inner support structure from shining through, and, on the other hand, irregularities are concealed better by multiple outer walls.

Z-seam Settings:

The "z-seam" is the vertical line that results from material deposits at the starting points of the individual layers. Position this seam at an inconspicuous place of the print (e.g., in the back area, or at a bend) by specifying the x-y-coordinates in the slicing settings. You can also use the "Hide z-seam" setting (slicing software: Cura). If you do not start the z-seam at an identical x-y-position, but rather "random," you will not see a defined z-seam in the printed object, but most likely so-called "blobs" or "zits." Nevertheless, you should decide for yourself which setting will give you a better print result.

Proper Print Temperature:

An appropriate printing temperature is the be-all and end-all of 3D printing. It is obvious that an inadequate low or high print temperature will ruin the best slicing settings. There are two effective ways: 1) Follow the filament manufacturer's temperature specifications (but this is usually only a range) and 2) print a so-called "temperature tower," which you can download from thingiverse.com. In the "gcodes" of these files, it is specified that the temperature of the nozzle is increased or decreased as the printing progresses. Thus, areas are created with varying print temperatures in one print job and can be compared 1:1. Simply choose the best temperature after a subjective quality assessment.

Smoothing the Surface:

Applying Cura's smoothing function ("Ironing") is a good way to refine the surface of a printed object. With this function, you can create very high-quality surfaces, which you will hardly see in FDM 3D printing.

Final Tips:

If you would like to learn more about how the numerous parameters in your slicing software affect the print quality, it is recommended to use the function "Setting per object" (Cura) positioned in the menu bar (left side). With this function, you can apply different settings for each object and thus print more versions of an object using different settings in one print job. In this way, you can compare the individual settings. When you have finally found the profile of your 3D printing dreams, don't forget to save these settings and make a backup copy of the profile. For those who are still searching, there is a 3D printing profile in the appendix of this book, which should provide high-quality prints (best to use with the CR-10 3D Printer).

Thank you very much for buying this book! If you liked it, it would be very helpful for me – and for other "Warping" and "Layer Splitting" plagued ones – if you would write a short review of this book.

Thank you very kindly for your support!

14 Bonus Material: Optimized Slicing Profile to use with Cura

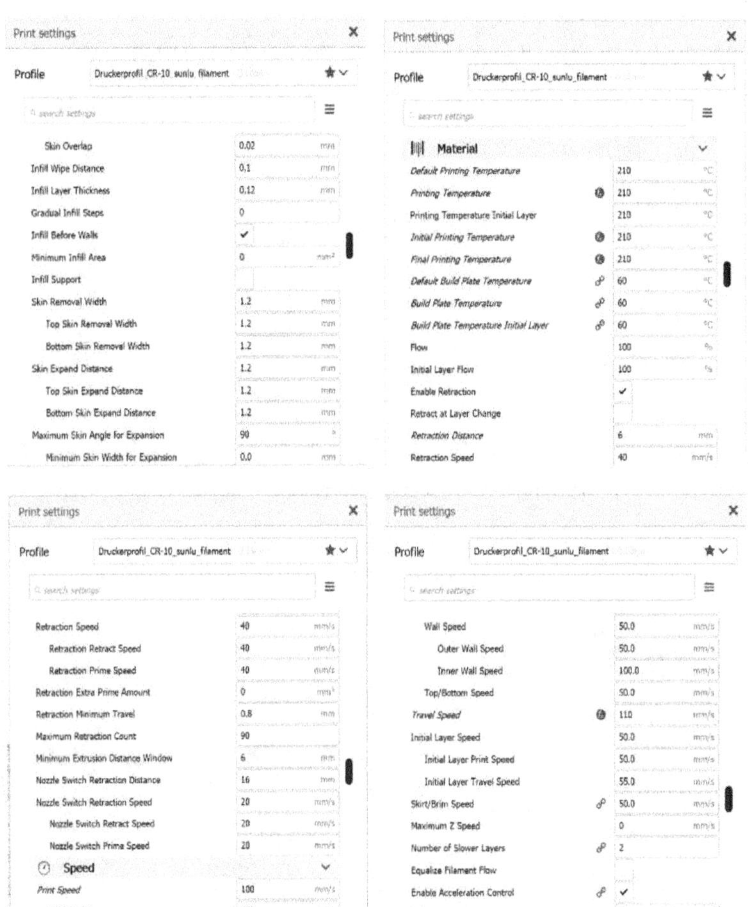

Print settings
Profile   Druckerprofil_CR-10_sunlu_filament
Skin Overlap   0.02 mm
Infill Wipe Distance   0.1 mm
Infill Layer Thickness   0.12 mm
Gradual Infill Steps   0
Infill Before Walls   ✓
Minimum Infill Area   0 mm²
Infill Support
Skin Removal Width   1.2 mm
Top Skin Removal Width   1.2 mm
Bottom Skin Removal Width   1.2 mm
Skin Expand Distance   1.2 mm
Top Skin Expand Distance   1.2 mm
Bottom Skin Expand Distance   1.2 mm
Maximum Skin Angle for Expansion   90 °
Minimum Skin Width for Expansion   0.0 mm
Print settings
Profile   Druckerprofil_CR-10_sunlu_filament
Material
Default Printing Temperature   210 °C
Printing Temperature   210 °C
Printing Temperature Initial Layer   210 °C
Initial Printing Temperature   210 °C
Final Printing Temperature   210 °C
Default Build Plate Temperature   60 °C
Build Plate Temperature   60 °C
Build Plate Temperature Initial Layer   60 °C
Flow   100 %
Initial Layer Flow   100 %
Enable Retraction   ✓
Retract at Layer Change
Retraction Distance   6 mm
Retraction Speed   40 mm/s
Print settings
Profile   Druckerprofil_CR-10_sunlu_filament
Retraction Speed   40 mm/s
Retraction Retract Speed   40 mm/s
Retraction Prime Speed   40 mm/s
Retraction Extra Prime Amount   0 mm³
Retraction Minimum Travel   0.8 mm
Maximum Retraction Count   90
Minimum Extrusion Distance Window   6 mm
Nozzle Switch Retraction Distance   16 mm
Nozzle Switch Retraction Speed   20 mm/s
Nozzle Switch Retract Speed   20 mm/s
Nozzle Switch Prime Speed   20 mm/s
Speed
Print Speed   100 mm/s
Infill Speed   100 mm/s
Wall Speed   50.0 mm/s
Print settings
Profile   Druckerprofil_CR-10_sunlu_filament
Wall Speed   50.0 mm/s
Outer Wall Speed   50.0 mm/s
Inner Wall Speed   100.0 mm/s
Top/Bottom Speed   50.0 mm/s
Travel Speed   110 mm/s
Initial Layer Speed   50.0 mm/s
Initial Layer Print Speed   50.0 mm/s
Initial Layer Travel Speed   55.0 mm/s
Skirt/Brim Speed   50.0 mm/s
Maximum Z Speed   0 mm/s
Number of Slower Layers   2
Equalize Filament Flow
Enable Acceleration Control   ✓
Print Acceleration   500 mm/s²
Infill Acceleration   500 mm/s²

Books on topics you might also like

All books are available online on the usual sales platforms. It's best to just search for the title, or feel free to visit my author page. Some of the books may not be published yet and will be released or found soon. Take a look at the books of your choice and your copy as e-book or paperback!

3D Printing:

CAD, FEM, CAM (3D Object Creation, Design, Simulation):

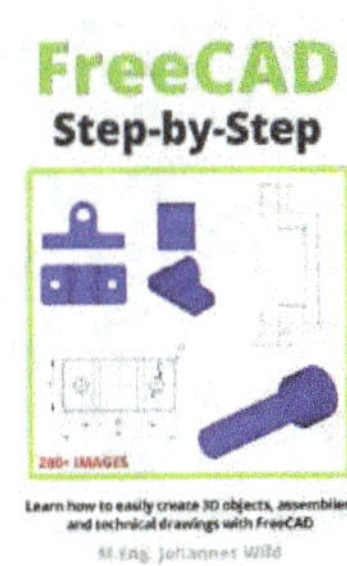

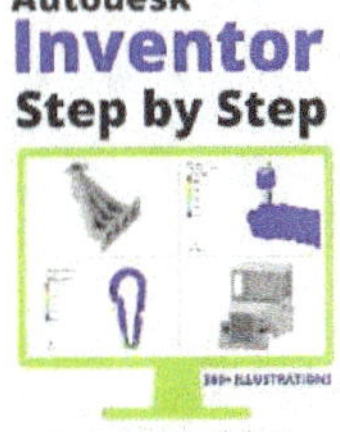

Electrical Engineering:

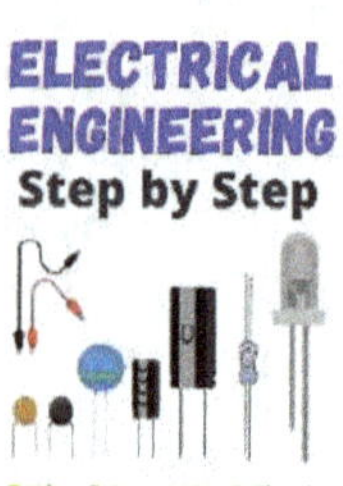

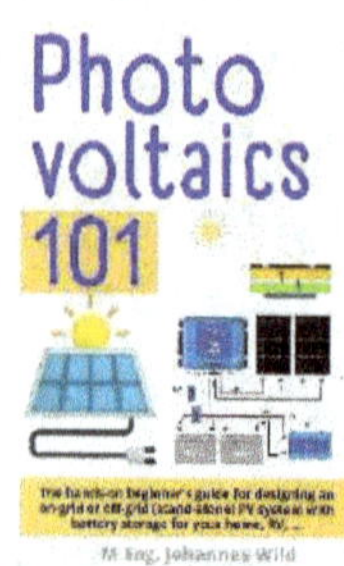

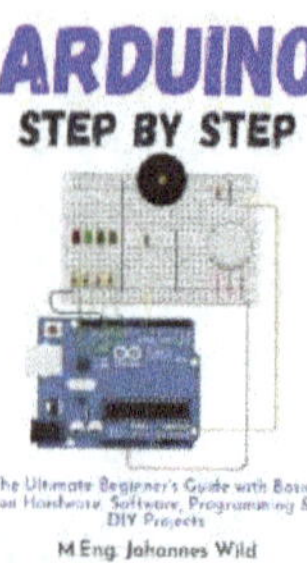

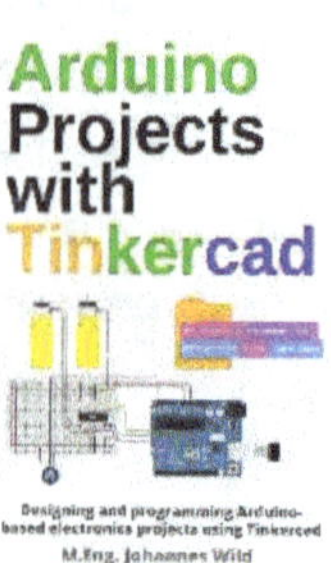

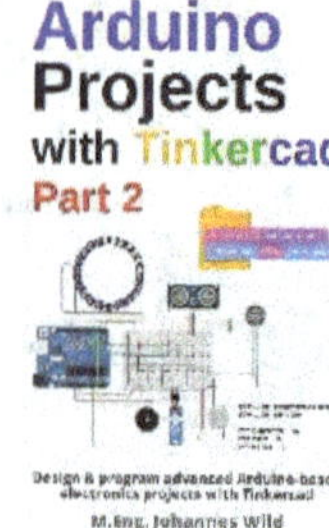

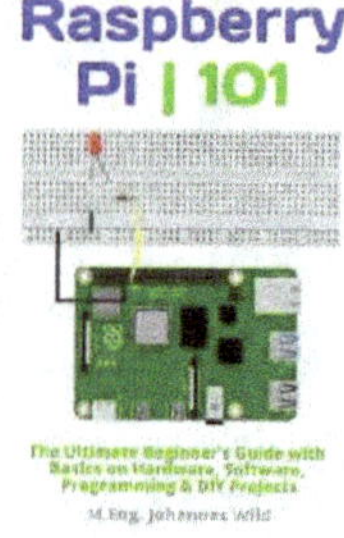

Programming and other Software:

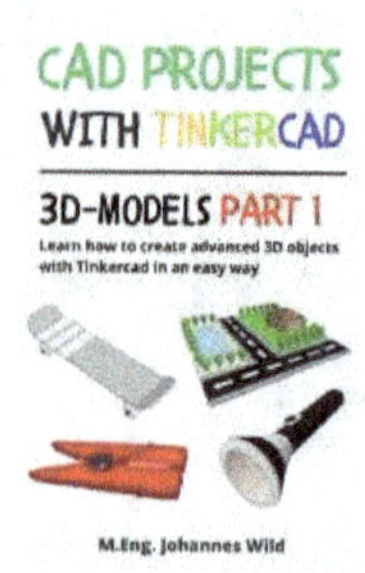

There are also identical video courses for these books:

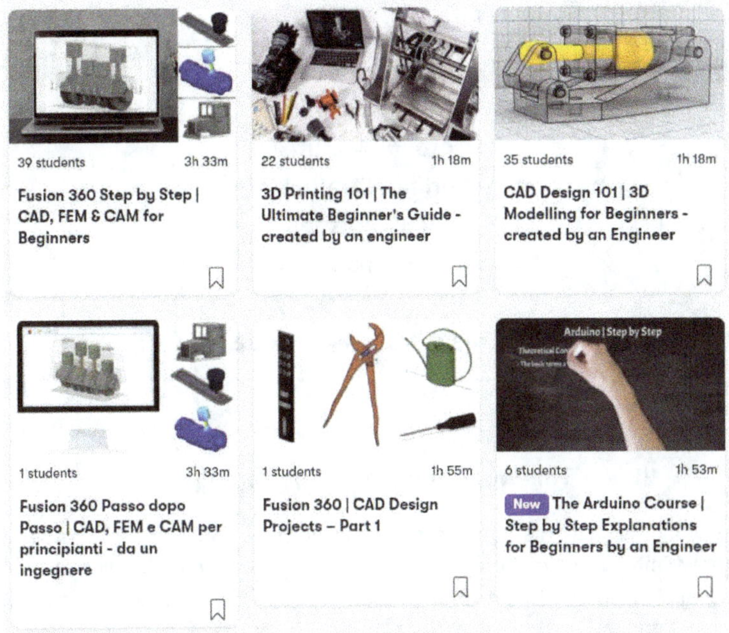

They are hosted on the learning website: skillshare.com

Be sure to use my following friends & family referral link to get a month of membership for free !

(I will get a little bonus if you choose to stay, so we will be both 😊 Thanks in advance!)

https://www.skillshare.com/r/profile/Johannes-Wild/854541251

It is best to copy the link in your browser to access the free month !

Sign up today and deepen your knowledge!

Imprint of the author / publisher

© 2023

Johannes Wild
c/o RA Matutis
Berliner Straße 57
14467 Potsdam
Germany

Email: 3dtech@gmx.de

All information contained in this book has been compiled to the best of our knowledge and has been carefully checked. However, this book is for educational purposes only and does not constitute a recommendation for action. In particular, no warranty or liability is given by the author and publisher for the use or non-use of any information in this book. Trademarks and other rights cited in this book remain the sole property of their respective authors or rights holders.

Thank you so much for choosing this book!